Mobile Disruption

Mobile Diruption
The Technologies and Applications Driving the Mobile Internet

Jeffrey L. Funk

A JOHN WILEY & SONS, INC., PUBLICATION

Published by John Wiley & Sons, Inc., Hoboken, New Jersey.
Published simultaneously in Canada.

For general information on our other products and services please contact our Customer Care Department within the U.S. at 877-762-2974, outside the U.S. at 317-572-3993 or fax 317-572-4002.

Wiley also publishes its books in a variety of electronic formats. Some content that appears in print, however, may not be available in electronic format.

Library of Congress Cataloging-in-Publication:

Funk, Jeffrey L.
    Mobile disruption : the technologies and applications driving
the mobile Internet / Jeffrey L. Funk.
        p. cm.
    ISBN 0-471-51122-6 (Cloth)
  1. Wireless Internet.   2. Wireless communication systems.   I. Title.
    TK5103.4885 .F86 2004
    004.67'8—dc22
                                                    2003020805

Printed in the United States of America.

10 9 8 7 6 5 4 3 2 1

To Emi

Acknowledgements

It is impossible for me to mention all of the people who have helped me with this book. Almost 150 people from almost as many firms provided with me information and valuable insights into the mobile Internet. Some firms, such as NTT DoCoMo, J-Phone, Index, Giga Networks, Tsutaya Online, Jeansmate, Asahi TV, and Toshiba granted me multiple interviews. Richard Gatarski, Daniel Scuka, and Jari Veijalainen read early drafts and provided me with comments. Many more people provided me with comments in my presentations of this material in a number of academic, consulting, and professional venues. Of course any errors are solely my responsibility.

Other people provided me with valuable financial, intellectual, and emotional support. The mobile Innovation Research Program at Hitotsubashi University provided me with financial support through grants from Fujitsu, J-Phone, KDDI, NEC, and NTT DoCoMo. Many professors including Akira Takeishi, Seichi Yonekura, and Yaichi Aoshima provided me with intellectual and emotional support. Norika Morimoto translated most of the Word Documents into Page Maker in order to create a "camera-ready copy." Danielle Lacourciere, Kirsten Rohstedt, and Val Moliere of John Wiley & Sons made the book possible. And of course my wife put up with my long working hours and business trips.

CONTENTS

Chapter 1

The Next Disruption

If you want to see the future of the mobile Internet, just empty your pockets, shoulder bags, and briefcases. In Japan, mobile phones are already being used as portable entertainment players, cameras, membership and loyalty cards, guidebooks, maps, tickets, watches, and devices for accessing everything from news to corporate databases. Within a few years, this list will likely include train and bus passes, credit and debit cards, keys, identification, and even money. It is also likely that the mobile phone will eventually replace many of these items. Mobile phones are already being accused of driving down the sales of watches and digital cameras, and the list will continue to get longer.

Clearly these changes will have a dramatic effect on our lives and on the firms that either make or utilize these products. For example, the ability to access corporate databases from phones will enable a whole class of workers to rely less on their memory and more on their company's and even world's knowledge. Maintenance, sales, construction, transportation, police, taxicab drivers, and many other workers spend a great deal of time away from offices. Thus, when they need information, they must either have it in their heads or stop what they are doing and make a phone call or even return to the office.

These changes will not happen overnight since networks, even if they are virtual networks, aren't created in a day. Transportation networks have been evolving since humans first appeared more than one million years ago, and even today's highways reflect thousands of small improvements. The same can be said of power and telephone networks and even the Internet where thousands of complementary innovations were needed for these networks to reach where they are today. Virtual networks like clothing manufacturers in Italy and IT firms in Silicon Valley are similar in that complementary innovations in social infrastructure have played critical roles.

The mobile Internet is no exception. Providing information on phones requires more than just new network technologies like packet-based infrastructure and high-speed data services; it also requires new phones, contents, components, and software. In particular, phones must be transformed from a voice-based to a multiple-media device. The old adage about which came first, the chicken or the egg, is more relevant than ever. As phone manufacturers wait for the successful contents and services to become available, many of the firms that have the capability to supply these contents and services are waiting for the relevant phones to emerge.

This book uses the concepts of network effects and disruptive technologies to describe how the mobile phone is becoming a portable entertainment player, a new marketing tool for retailers and manufacturers, a multichannel shopping device, a navigation tool, a new type of ticket and money, and a new mobile Intranet device. Network effects explain the phenomena of the chicken and egg. Disruptive technologies represent a class of new technologies whose initial performance limitations cause unexpected customers and applications to emerge. Each chapter addresses a different application, and they use these concepts plus competition in current markets, of which the largest is the Japanese market, to analyze the evolution of services, user needs, and competition and how key technological trajectories are changing the design tradeoffs and competition.

NETWORK EFFECTS

Economists use the term *network effects* or *positive feedback* to describe the phenomenon of the chicken and egg[1]. The value of a product that exhibits network effects is a function of the number of users, of which there are direct and indirect effects. The telephone, facsimile, and e-mail exhibit direct effects. The more people you can call or send e-mail, the greater the value of these products. Typically, a critical mass of users is needed to initially create positive feedback in the system; the picture phone is one example of a product that never created a critical mass of users.

Indirect effects involve networks that require the use of complementary products. Computer networks have hardware and software, television broadcasting involves programming and receivers, and automobiles require gasoline stations. The more software there is for computer hardware, the more people want to buy that hardware. On the other hand, a critical mass for digital television or electric vehicles has not yet emerged. Consumers are waiting for broadcasters to offer digital programming and the broadcasters are waiting for the consumers to buy the digital televisions.

The existence of these network effects and their required critical mass increase the importance of identifying those customers who will put up with the limitations of the new technology. The initial performance of all new technologies is far inferior to their potential performance. However, the existence of network effects and the need for a critical mass increase the importance of these initial customers and their so-called killer applications. Without them it is difficult to convince all the producers of complementary products to produce appropriate products that work together.

The heavy initial investments that are required in most network industries also make the identification of these initial customers and their applications important. The stock market typically requires firms to demonstrate profitability in these new networks quite early. True, they may give firms a few years to become profitable, but eventually profitability must be produced as many dotcoms have found the hard way.

If the customers for the new technology are not much different from the lead customers of the old technology, finding the appropriate customers may not be very difficult. Oftentimes the major limitation of new technology is merely its high price. In this case, the challenge is to find those customers who have deep pockets as was the case in the early days of integrated circuits in the United States. The military provided the initial market for integrated circuits and thus provided the means to develop smaller, denser, and cheaper integrated circuits that found their way into other products like computers, as well as develop consumer electronics through increasing returns and network effects.

Problems occur when the most appropriate initial customers are different from the existing customers. Clayton Christensen coined the term *disruptive technologies* to describe these kinds of technologies[2].

DISRUPTIVE TECHNOLOGIES

Disruptive technologies improve some aspects of performance while reducing others, thus causing a new set of customers to be the first users. These new customers may be completely different from the existing customers of the old technology or merely different from the lead customers in the old technology. Even if they are only different from the lead customers, this can cause major problems for incumbents since their managerial processes are often organized to support these lead customers.

Put more bluntly, firms that do not create systems to support their lead customers rarely stay in business a long time. Successful firms study the needs of these lead customers, show them new prototypes, and listen carefully to their feedback. But when the new technology is more appropriate for a new set of customers, this knowledge of and focus on existing customers becomes a weakness. This is why Christensen calls his book on disruptive technologies *The Innovators Dilemma*. With disruptive technologies, firms must simultaneously support their existing customers customers, while they search for new ones. This is particularly difficult when the margins for the new customers are smaller than those with the existing customers since incumbents would prefer to move upstream to those customers who have higher margins than downstream to those who provide lower margins.

A second characteristic of disruptive technologies is that technical improvements often cause the market for the new technology to become larger than the market for the old technology. The new market may grow faster than the market for the old technology, and the new technology may become applicable to the existing customers of the old technology. This even occurs when the old and new technology have similar rates of improvement, which is most often the case. The problem is that the old technology often ends up providing more performance than is demanded by customers on a single dimension. When this occurs, the focus of competition often changes to price or another dimension of performance where the new technology often has an advantage.

Several examples of disruptive technologies, of which the computer industry is the easiest to understand and the most relevant to the mobile Internet, are shown in Table 1.1. As shown in Figure 1.1, each new generation of computing technology has offered significantly lower processing performance in return for lower costs and other advantages. These lower processing speeds have caused each new generation of computing technology to appeal to a new set of users; in turn this has caused the incumbent firms to largely ignore the next generation of technology.

The PC is the most famous example because it first created an entirely new computing market, it has subsequently been used in place of

TABLE 1.1. Examples of Disruptive Technologies in the Computer and Electronics Industry

New Technology	Old Technology	Potential Advantage	Disruptive Nature
Transistor Radios	Vacuum Tubes	Smaller, cheaper	Poor sound quality
Transistor TVs	Vacuum Tubes	Cheaper and more reliable	Low voltages and thus small sizes
Smaller disk drives	Larger disk drives	Smaller, cheaper & more reliable	Less memory capacity
Mini-computers	Mainframes	Cheaper, easier to modify	Less processing capability
PCs	Mini-computers	Smaller & cheaper	Poor processing capability
PDAs	PCs	Smaller & cheaper	Poor processing & input-output capability
PC Internet	Print media	Greater richness, reach	Initially low richness
Mobile Internet	Fixed-line Internet	Portability	Small size

FIGURE 1.1. Four Generations of Processing Speed

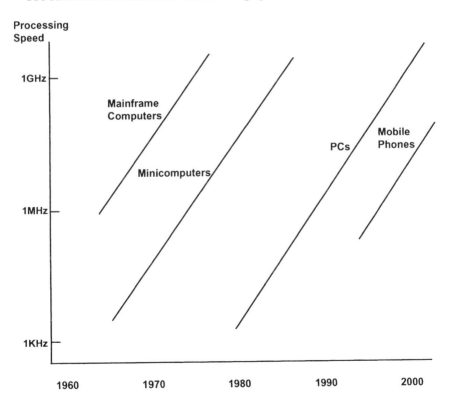

mainframes and mini-computers, and IBM's comments about the PC in the late 1970s have been widely quoted. IBM reportedly called the first users "hobbyists" and said the technology had no relevance for its customers[3]. By the time IBM recognized the importance of the PC, other firms had already developed more capabilities in writing software for the performance-restricted PC than IBM had and thus IBM had to purchase the software; Microsoft was the lucky selection. Critics of IBM should keep in mind that it has done far better than any of its mainframe and mini-computer competitors like Burroughs, Remington, Honeywell, DEC, Wang, and Prime.

PC manufacturers and Microsoft have also not done well in the PDA market. The hardware incumbents focused on the needs of lead users of PCs, who were concerned about processing speeds, memory, and word processing capabilities, and tried to sell them so-called hand-held computers. Similarly, Microsoft has tried to sell a product that is compatible with the old technology when doing this was both initially difficult and apparently not very important to PDA users. On the other hand, Palm found a set of users who were willing to sacrifice these capabilities for the ability to manage their schedules and addresses and write memos[4].

The mobile Internet is also a disruptive technology when compared to the PC Internet. Mobile phones have higher portability but inferior displays, keyboards, and processing and memory capability than PCs. This makes the mobile Internet most appropriate for those users who value portability (users who spend a large amount of time away from offices or home) over display, keyboard, and processing capability (users who are less interested in complex applications). The large number of young mobile Internet users in Japan and young SMS (short message services) users in Europe and Asia supports the notion that young people are these users.

In general, young people are more mobile than older people due to better health, less access to offices (high school and college students rarely have offices), and less need to be in their homes in the evenings (fewer children) than older people. Due to their lower emphasis on complex business applications and their higher propensity for trying new things, young people are also probably more willing to put up with the smaller displays, keyboards, and processing capability of phones when compared to older people. But it is the young people who place the most emphasis on fashion, coolness, and being seen in the right places (which makes it hard for them to be at home with their PC) who appear to be the largest users of the mobile Internet.

The second characteristic of disruptive technologies is more uncertain, but it is at the heart of the controversy over the future of the mobile Internet. Two key trajectories are display size and processing capability. In addition to the 5 - 10% increases in display size each year that come from reducing display thickness through currently understood technologies, new technologies like light-emitting polymers provide paper-thin displays that can be rolled and folded[5]. Such technologies may make it possible to dramatically increase the size of the phone display in the future without increasing the size or weight of the phone; of course, predicting when this will occur is a more difficult question.

Predicting the future of application processors is much easier since the roughly doubling of processor and other chip capability every 18 months has been going on for almost 40 years. The lower prices and greater importance of low power consumption requires the use of different processors in the phones than are used in PCs. Mobile phones with processing speeds greater than 100 MHz first appeared in 2002 as compared to such PCs in 1993, or a difference of about nine years.

One question is how improved application processors (both faster speeds and lower power consumption) will lead to a better user interface. Client-based programming languages like Java are already changing the user interface in the Japanese mobile Internet and the 100-MHz processor can handle single-word voice recognition, speech synthesis, and simple 3D imaging techniques. Future improvements in speed and power consumption will increase the sophistication of 3D imaging techniques, improve the reliability of single-word and make multiword voice recognition possible, and at some point in the future make new interfaces like virtual reality or holograms possible.

These developments may cause PDAs to play the role of workstations in the mobile Internet. Workstations have always had more processing capabilities than PCs, but their unit sales are currently less than 1% of the number in the PC market. Furthermore, their performance advantage has become less important as PC performance has improved. Similar things appear to be happening in the mobile Internet where more than 30 times the number of mobile phones were sold in 2002 as PDAs.

But we are getting ahead of ourselves. There are additional issues with respect to the concept of disruptive technologies that need to be described. In particular, many readers are probably wondering what the issue of disruptiveness has to do with the greater success of the mobile Internet in Japan and Korea than elsewhere. The Japanese mobile Internet probably represents more than 75% of the global market for mobile Internet services, with Korea a strong second. As of the end of 2002, there were almost 35 million Koreans[6] and more than 70 million Japanese using their phones to access a limited form of the Internet; the Japanese were spending more than $15 a month in subscription, packet, and content charges (see Table 1.2).

One reason for Japan's greater success is that NTT DoCoMo created its i-mode service using an approach similar to that used by other successful incumbents with disruptive technologies. NTT DoCoMo created a separate organization and hired outsiders like Mari Matsunaga and Takeshi Natsuno, who had experience in other industries. They focused on consumers as opposed to business users and created a micro-payment system in which NTT DoCoMo collects charges for content providers; the micro-payment system (see Figure 1.2) made it easier for entertainment and other consumer-oriented content providers to collect revenues. But an additional reason why NTT DoCoMo focused on consumers as opposed to business users bring us back to the concept of disruptive technologies and the different paths followed by Japanese and Western markets.

DISRUPTIVE TO WHOM?

One reason why existing firms respond to a new technologies in different ways is that they have traversed different paths in particular with different customers. Different customers cause firms to develop different business models, pursue different forms of improvements, and develop different internal and external communication channels. Thus, the disruptiveness of a new technology to an existing firm can be measured by the degree to which the new technology is appropriate for a firm's existing customers.

Table 1.3 lists some technologies that have been more disruptive for U.S. firms than for Japanese firms because the Japanese electronic firms were more focused on the first applications for these new technologies than were the U.S. electronics firms. For example, CMOS (complementary metal oxide semiconductors) technology was much more disruptive for U.S. electronic manufacturers than for Japanese ones because the first applications were in products the Japanese electronic firms were already aggressively pursuing like calculators and watches. The emphasis by Japanese firms on calculators and watches caused them to also play a great deal of emphasis on the development of semiconductors that have low power consumption like CMOS while U.S. semiconductor manufacturers were focused on the computer market and thus high-speed devices like bipolar semiconductors.

Japan's early involvement in these low power consumption devices

TABLE 1.2. Size of Japanese Mobile Internet in 2002   (Billion of Yen)

Type of Market	Firm or Market	Size of Market
Services (data charges)	NTT DoCoMo	697.7
	KDDI	138.7
	J-Phone	185.7
Contents	Total	150
	Ringing tones	80
	Screen savers	25-30
	Games	15-20
	Other entertainment	5-10
	Other contents	10-15
Shopping		30-40

Source: Company documents

FIGURE 1.2. Overview of Micropayment System

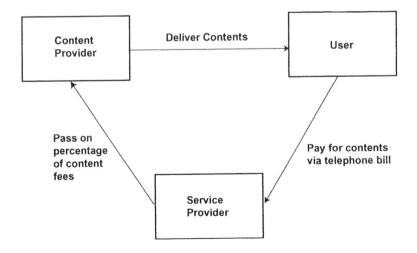

TABLE 1.3. Examples of Technologies that were more Disruptive to US than Japanese
Firms

New Technology	Old Technology	Potential Advantage	Disruptive Nature	Initial Applications
CMOS chips	MOS and Bipolar chips	Lower power	Slower	Portable calculators and watches
LCDs	Vacuum Tubes	Cheaper	Low yields	Portable calculators and watches
Solar cells	Batteries/ Fossil fuels	Renewable resource	Low efficiencies & power output	Calculators
CCDs	Vidicon Tubes	Cheap & portable	Poor resolution	Facsimiles & video cameras

not only helped them dominate the calculator market but also assisted in their temporary domination of the global semiconductor market as concerns about heat dissipation in semiconductors caused MOS and later CMOS to replace bipolar technology in memory chips[7]. Similarly, Sharp and Seiko commercialized LCDs (liquid crystal displays) faster than RCA (who was the first firm to develop LCDs) since this new technology was initially appropriate for watches and calculators and not RCA's key market, televisions where there were insufficient yields for large displays (this has finally begun to change). Sanyo and Sharp have commercialized solar cells faster than U.S. power generation suppliers since solar cells were initially more appropriate for calculators than for power generators[8].

Nevertheless, it was another disruptive technology, the PC, which caused U.S. firms to regain domination of the semiconductor industry. Intel developed the microprocessor through a chance order from a Japanese producer of calculators and its decision to retain the rights to the technology when the customer wanted to renegotiate the price. The microprocessor was much less disruptive for U.S. firms than for Japanese firms because it was first used in a number of industrial applications where U.S firms were the leaders. As PCs diffused, the microprocessor and other key parts in the PC began driving the semiconductor market, thus replacing consumer electronics as the key driver in this market.

The mobile Internet is more disruptive for Western firms than for

Japanese and Korean firms because the Western firms focused on business users more than Japanese and Korean firms when the target market for the mobile Internet appears to be young users. Of course all mobile phone markets initially started with business users in the 1980s. But as the mobile phone market expanded from business to consumers in Europe, the United States, and many parts of Asia in the late 1990s, Japanese and Korean firms began shifting their emphasis to consumers while European and U.S. mobile phone manufacturers and service providers remained focused on business users, particularly for their mobile Internet services.

One reason why European and U.S. mobile service providers remained focused on business users is due to the greater importance of both corporate users and roaming (roaming is primarily used by business users) in Europe and the United States. Part of the reason for the greater importance of roaming outside of Japan and Korea is the success of GSM. Although it has not been adopted by either Japan or Korea, more than 170 other countries have adopted GSM. And the many GSM committees and conferences have an enormous impact not only on the technologies that are adopted by the world's mobile service providers, but they also impact on business models as WAP grew out of the GSM world. These committees and conferences have contributed to a fairly unified viewpoint on the importance of business users and applications for the mobile Internet in the GSM and WAP worlds.

A second reason is that Japanese and Korean firms began to shift their emphasis toward the consumer market through their payment of higher phone subsidies to retail outlets to sign up new subscribers. These high subsidies have enabled Japanese and Korean manufacturers to introduce much more technology (including mobile Internet technology) into phones than their European counterparts and still produce phones that are basically free to the final consumer within six months of their release. This also caused the Japanese and Korean phone markets to collapse into one rather inexpensive price segment where manufacturers compete to introduce the latest technologies like better displays and internal cameras. One the other hand, European and U.S. mobile phone markets have multiple price segments where manufacturers continue to target business users with their high-end phones, which are currently mobile Internet-compatible phones.

The lower emphasis placed on business users by Japanese and Korean firms made it easier for them to ignore business users and focus on general consumers in the mobile Internet. This is one reason why NTT DoCoMo included a micro-payment system and entertainment contents in its i-mode service. The success of entertainment contents such as ringing tones, screen savers, and games created a critical mass of users, which caused the number of content providers, compatible phones, and users to explode within one year of the start of i-mode services in February 1999. Furthermore, the success of i-mode also caused the other two Japanese service providers to copy and in some ways improve NTT DoCoMo's business model.

NETWORK EFFECTS AND COMMUNICATION CHANNELS

Positive feedback and network effects also drive regional specialization. Firms tend to locate near customers and producers of complementary products, and Silicon Valley is the most famous example. Furthermore, regional specialization shapes and is shaped by communication channels. Thus, a successful technology will create positive feedback in the system as firms share information about the ways in which the technology can be successfully applied. On the other hand, a technology that is disruptive to a specific region can cause the opposite to occur; firms will ignore the technology and discount the success of the technology in another region as an anomaly that can be ignored.

For example, the transistor radio and television were both disruptive technologies. The disruptiveness of the former was due to the poor frequency characteristics of early transistors, which caused poor sound quality and led to a bad quality reputation for Japanese manufacturers. The disruptiveness of the latter was due to the low voltage capability of early transistors, which prevented the initial manufacture of large screen televisions. These problems caused incumbent manufacturers of radios and televisions, both U.S. and European manufacturers, to initially ignore the transistor and it was Sony, then a new venture, which made the strongest investments in transistors and became the leader in transistor radios and televisions[9].

Sony's early success in transistor radio and later transistor televisions caused many other Japanese manufacturers such as Toshiba and Hitachi[10] to release these products. On the other hand, many U.S. and European firms derided the poor quality of Japanese transistor radios and televisions throughout the 1960s and were slow to use transistors and ICs in their radios and televisions. Even as late as the 1970s, many U.S. firms were unable to recognize the important role that Japan's early application of transistors had played a critical role in their success. As a result, Japanese manufacturers followed a distinctly consumer electronics path in the electronics industry while U.S. firms concentrated on the military (an option not available to the Japanese market) and computer markets.

A similar thing is now occurring in the mobile Internet. The early success in Japan has caused positive feedback to emerge among content providers, phone manufacturers, technology suppliers, and corporate users while the lack of success in the United States has caused a different viewpoint to emerge. Like the early days of the consumer electronics industries, many U.S. business leaders, academics, and newspapers argue that the success of the mobile Internet in Japan is due to demand conditions that are unique to Japan. For example, many cite the large role played by public transportation in Japan in spite of the fact that mobile Internet usage in Japan is independent of region, and rural areas that resemble many of the Western states of the United States in terms of population density have similar penetration and usage rates as Tokyo.

Others argue that Japan's low penetration rate has been the reason for Japan's early success in the mobile Internet[11]. For example, well-respected technological commentators such as Andrew Seybold (editor of *Forbes/Andrew Seybold's Wireless Outlook)* has repeatedly made this argument including a stinging criticism of i-mode - "i-mode is a cultural success - not a wireless success" - and a prediction that PDAs would become the dominant form of wireless access[12]. Technological commentators and Andrew Seybold would be wise to ask why the United States was one of the first countries to successfully introduce mobile phones, hand-held computers, and portable tape players when the United States already had one of the highest penetration rates for fixed-line phones, PCs, and home entertainment systems. They might even ask why they themselves own most of these products.

Still others argue that W-LAN will be the network for the mobile Internet, with Americans accessing the Internet from their laptops in coffee shops and other places. This argument ignores the fact that the network and the terminal are two different issues. Even if W-LAN replaces mobile phone networks, mobile phones can and most likely will be used in these networks. This is one reason why mobile service providers from all over the world are trying to offer both W-LAN and third-generation mobile services so that mobile phone users can easily switch between the two networks.

Europeans are caught somewhere in the middle between the Japanese and U.S. viewpoints. Like the United States, mobile Internet services in Europe, called WAP, initially flopped due to their emphasis on expensive phones and complex business application, of which the latter were too difficult to implement on the tiny screens and keyboards found on mobile phones. Instead, SMS, which is basically a non-Internet-based mail service, has been far more successful than the mobile Internet services in Europe, Asia, and even in the United States partly because contents and new phones with large displays were not needed to create a critical mass of users. SMS first succeeded in Scandinavia, which had the highest PC Internet usage in Europe at that time (see Figure 1.3). Furthermore, these services contained an easy way to achieve revenue sharing between service and content providers thus causing many young people to download ringing tones and screen savers with them; this suggests that culture has little to do with the success of these entertainment contents in Japan.

Several European and U.S. service provideers most notably Vodafone have learned from the success of SMS in Europe and Japan. Vodafone has learned about the Japanese mobile Internet through its Japanese subsidiary, J-Phone; Vodafone introduced Vodafone Live! in late 2002. The service combines WAP, MMS, and other technologies. The number of subscribers for the Vodafone Live Service passed the two million mark in August 2003 or about three times the number of i-mode

FIGURE 1.3. SMS Usage Versus PC Internet Penetration

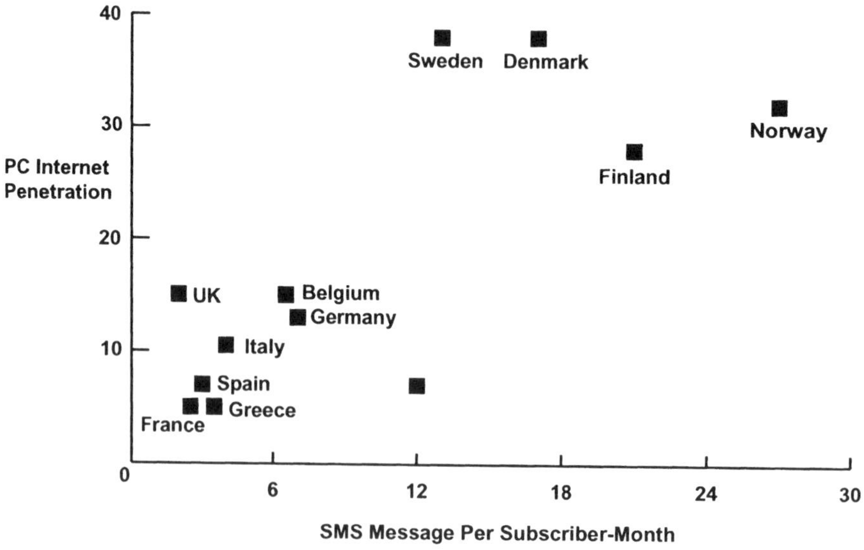

Source: PC Internet Penetration; US Internet Council
        SMS Message Per Subscriber-Month; GSM Association

subscribers in Europe. Like i-mode, ringing tones and screen savers are the most popular services[13].

The United States would be wise to learn from the success of SMS and Vodafone Live in Europe. Usage of SMS in the United States has lagged Europe due to the compatibility problems between different digital standards, but growth started as soon as these problems were solved in mid-2002. U.S. firms should aggressively implement mobile Internet services since they are superior to SMS. Mobile Internet services provide browsing and Internet mail; the latter enables firms to include URLs in mail messages, something that is not possible with SMS. Like the PC Internet, the mobile Internet has the potential to change our personal and business lives while SMS is merely an intermediate technology that will someday be seen in the same light as France's Minitel or other pre-Internet proprietary information systems.

COMPETITION IN THE MOBILE INTERNET

This book uses the concepts of network effects and disruptive technologies to explain how the market and competition in the mobile Internet is currently and may evolve in the future. Chapter 2 describes the origins and early evolution of the mobile Internet using a model of industry formation. The intersection between several key technological trajectories such as improved phones and phone networks, and the digitalization of content made the mobile Internet possible and it was young people downloading entertainment related contents that caused the Japanese market to initially grow. The initial growth led to technological innovation, user learning, and an expansion in the applications.

Chapter 3 describes the sub-trajectories that emerged from the growth in the Japanese mobile Internet of which one important sub-trajectory is the processing capability of phones, which already equals that of PCs released in 1994. Increased processing speeds and lower battery consumption enable more client-side processing; in the short run this will increase the size and complexity of programming languages like Java while in the long run the increased processing power will enable more sophisticated types of user interfaces to emerge that further compensate for the small screens and keyboards of mobile phones.

Chapters 4 through 9 use the concepts of network effects, disruptive technologies, and the sub-trajectories to describe and forecast growth in six key contents/applications, and the existing and future competition in them. It is based on published information from both Japanese and English sources and interviews with more than 150 managers invovled in the mobile Internet in Japan and to a lesser extent elsewhere as both users and suppliers of the mobile Internet. I asked these managers about the current and future impact of the mobile Internet on their businesses with a focus on so-called "lead users." I used this information to describe various paths by which the mobile Internet may evolve in six contents/applications.

Chapter 4 describes the business models and technologies that are turning mobile phones into portable entertainment players, and it also discusses the opportunities this presents for new firms due to the disruptive nature of the mobile Internet. Higher network speeds and lower packet charges are making the downloading of music and videos possible. Increased processing power and larger Java programs are enabling content providers to turn screen savers into browsers and combine ringing tones, music, and video with these browser-enabled screen savers to create multimedia entertainment services that will redefine the portable entertainment market.

Chapter 5 describes the business models and technologies that are making the mobile phone the "new contact point for young customers[14]." Retail outlets and manufacturers are using the mobile Internet (and SMS outside of Japan) to send discount coupons, conduct surveys, offer free samples, and improve their brand image with young people. New technologies such as 2D bar codes, short-range infrared, Java, and devices that activate a phone's mail function offer additional ways for retailers to develop stronger relationships with customers, including the use of phones as mileage/point cards. The disruptive nature of the mobile Internet provides the most opportunities to those retailers and manufacturers who sell to young people and to the firms who supply these technical applications.

Chapter 6 describes the business models and technologies that are enabling the mobile phone to become a multichannel shopping device where users purchase products in combination with other media such as magazines and radio and TV programs. Already the disruptive nature of the mobile Internet has caused the successful products, customers, and selection methods to be different from that of the mobile Internet and these differences have led to the success of a new set of firms. These new firms, along with many incumbents, are now attempting to integrate mobile Internet services with other media such as magazines, radio programs, and television programs where larger and more complex Java programs, bar codes, cameras, and other technologies will drive this multichannel integration.

Chapter 7 describes the business models and technologies that are turning the phone into a portable navigation device. Phones are being used to choose and find destinations, and firms are integrating these B2C service with more sophisticated B2B services. Various technologies such as GPS make location-services possible, and technological improvements will continue to reduce the costs of these services. This chapter also uses the concept of disruptive technologies to explain how it is a new set of users that are driving these navigation services, both the GPS-based and the non-GPS-based services.

Chapter 8 describes the applications and technologies that are enabling mobile phones to be used as tickets and money, thus continuing the move from physical to electronic money that was started with credit cards 50 years ago. Key technologies include two-dimensional bar codes, infrared connections, and smart cards; smart cards are the early, but not necessarily, final leader. Smart cards are already used as transportation tickets and prepaid money in convenience stores, and it is somewhat simple to put these smart card functions in phones. While credit card companies initially led the efforts to develop contact cards, non-contact cards have faster processing times, which are needed for the ticket and convenience store applications where smart cards have done very well. Increased processing power (which enables biometrics) and network effects may enable a new set of winners to challenge credit card companies on their own turf.

Chapter 9 describes the applications and technologies that are making the mobile phone a key part of firm's mobile Intranets. It also uses the concept of disruptive technologies to describe how a new set of users is implementing mobile Intranets. Instead of firms using mobile phones to access existing ERP (Enterprise Resource Planning) and CRM (customer relationship management) systems, there is new a different set of users, many of whom are the sources of the key innovations. The early applications include delivery, construction, maintenance, and taxis, and the key technological trends include larger displays, increased processing power and network speeds, Java, and the effect of these on improving the phone's user interface.

The last chapter analyzes platform strategies for service, content, handset, and technology providers. While the popular press often describes this battle in terms of Nokia versus Microsoft, the issue is actually far more complex and interesting. The key interface is between the user and the contents, and one of the key trajectories is processing power. Improvements in processing speed and its effect on Java, 3D rendering techniques, and other technologies are expected to impact on the design of the mobile Internet and thus play a key role in these platform strategies. Potential winners range from heavyweights such as Microsoft, Nokia, Sun Microsystems, and Intel to small and relatively unknown Japanese firms like Cybird and Index.

NOTE

[1] For example, see Shapiro, C. and H. Varian, *Information Rules*, Harvard Business School Press, 1999.

[2] See Christensen, C., *The Innovator's Dilemma*, Boston: Harvard Business School Press, 1997.

[3] For example, see Carrol, P., *Big Blues: the unmaking of IBM*, NY: Crown Books, 1983.

[4] See for example, Butter, A. and D. Pogue, *Piloting Palm: The inside story of Palm, Handspring, and the birth of the billion-dollar handheld industry*. On the other hand, the increased processing speeds in PDAs is now making it possible for Microsoft to provide compatibility between PDAs and desktop computers while Palm has not effectively used network effects to create barriers to entry.

[5] For example, see the discussion of new display technologies from E Ink and Gyricon in Kharif, O., *Business Week*, June 18, 2002, Special Report on Emerging Technologies.

[6] "Mobile Communications in Korea; History and Future," Presented in May 28, 2002, http://dis.cnu.ac.kr/download/Yang_Seungtaik-Mobile_Communications_in_Korea.pdf

[7] Many researchers have argued that Japan's success in the semiconductor industry in the 1970s and 1980s came from their success in the consumer electronics industry and its role as a driver of MOS and CMOS technologies. For example, see Langlois, R. and E. Steinmueller, 2000, "Strategy and circumstance: the response of American Firms to Japanese competition in semiconductors, 1980-1985," *Strategic Management Journal*, Vol. 21, pp. 1163-1173.

[8] These three examples were found through an analysis of Johnstone's history of Japan's consumer electronics industry: *We were Burning: Japanese Entrepreneurs and the Forging of the Electronic Age*, NY: Basic Books, 1999.

[9] Christensen, C., T. Craig, S. Hart, "The great disruption" *Foreign Affairs*, March-April 2001, pp. 80-95.

[10] Johnstone, B., *We were Burning: Japanese Entrepreneurs and the Forging of the Electronic Age*, NY: Basic Books, 1999.

[11] For example, see Markoff, J., "New Economy; The Internet in Japan is riding a wireless wave, leaving phone-line companies in the dust," *New York Times*, August 14, 2000 and "I-mode success," *The Economist*, March 9, 2000.

[12] For example, see a summary of his comments in 2000 see: http://www.mobic.com/news/index.jsp and http://www.mobic.com/oldnews/2000/08/ntt_docomo.htm

[13] Clark, M., "Vodafone claims success with new service," *Yahoo News*, March 26, 2003.

[14] I am indebted to Jamie Cattell for the term - new contact point for customers.

Chapter 2

The Origins of the Mobile Internet

Mobile phone services were started in the United States, Japan, Europe, and other leading industrialized countries in the early 1980s. Phones were initially based on analog technology and digital services were introduced in the 1990s such that by the mid-1990s, most major industrialized countries had introduced digital services. By the end of 1999, most industrialized countries had penetration and growth rates that exceeded 30% and 20% per year, respectively, and the rapid diffusion of the mobile phone had become a global phenomenon with more than one billion mobile phone subscribers in the world by late 2002[1].

Furthermore, the growth pattern was similar over most parts of the world. In each country, it began with business users in their thirties and forties and gradually expanded such that in countries with penetration rates greater than 60%, most people between the ages of 15 and 65 own phones. As implied by the penetration rates shown in Table 2.1, mobile phones first diffused to young people in Scandinavia followed by Korea, the rest of Europe, Japan, and the United States.

Service providers in these countries began introducing mobile Internet services in 1999 with very different results. This chapter summarizes the evolution and performance of these services origins using a simple model of industry evolution. The purpose of this discussion is to understand such issues as 1) How did the mobile Internet initially grow in terms of users, services, contents, and phones in Japan? 2) How and why did the industry evolve differently in the United States and Europe? Table 2.2 summarizes lessons learned from the success of the

TABLE 2.1. Selected Data for Mobile Phone Markets

Item	Japan	Korea	Scandinavia	United States	Europe
Phone penetration					
(1999)	38%	50%	55%	29%	39%
(2001)	53%	68%	76%	44%	79%
Digital standard	PDC CDMA	CDMA	GSM	CDMA TDMA GSM	GSM
Roaming revenues	<1%	<1%	10%	11.4%	10%
Corporate users	10%		30%-40%		
Phone subsidies by service providers	$250-$350	$250-$350	$70-$160		$70-$380
PC Internet penetration (1999)	21%	27%	41%	42%	19%

Source: Mobile phone penetration data from (Mobile Communication International, 2000-2002), roaming revenues from CTIA's home page, Credit Suisse First Boston Securities, and author's interviews with firms, data on subsidies from interviews and Credit Suisse First Boston Securities, and Internet penetration data from the United States Internet Council (2000).

TABLE 2.2. Lessons Learned from the Mobile Internet and SMS Services

1. Young people are the initial users.
2. PC Internet penetration is unrelated to mobile Internet and SMS usage.
3. Public transportation is largely unrelated to mobile Internet and SMS usage.
4. Simple entertainment such as ringing tones and screen savers are initially the largest content markets.
5. Micro-payment systems are needed for effective entertainment contents to appear.
6. The success of ringing tones and screen savers is a global phenomenon and not a cultural issue.
7. The specific type of markup language in the mobile Internet is not critical.
8. A service provider's control of phone specifications is not necessarily critical.
9. A packet system is useful but not absolutely critical.
10. Service providers should not over hype mobile Internet services. They are not the equivalent of the PC Internet in your pocket.
11. The mobile Internet, whether it is based on WAP or c-HTML, is a superior technology to SMS.
12. The mobile Internet is not just a service; it is a new network industry that requires complementary innovations by service, content, and software providers, and manufacturers.

mobile Internet in Japan and Korea and SMS elsewhere.

MODEL OF INDUSTRY FORMATION

A model of industry formation is summarized in Figure 2.1 that draws on many descriptions of technological change[2]. Existing and to some extent new industries drive improvements in various technologies where it is the combination of these technologies in new and novel ways that lead to the formation of new industries. The speed with which these new technologies are effectively combined and commercialized depends on the design of products and services for a set of initial applications where the economic incentives are high. Experimentation by both producers and users leads to the identification of these initial applications and thus the formation of a new industry. Growth in these initial applications causes sub-trajectories, where competition in the new industry initially takes place, to emerge from the main technological trajectories. In turn, these sub-trajectories expand the market for the initial applications and cause a new set of applications to also emerge. Positive feedback and network effects play important roles in a) the initial interactions between products and users in the market, b) the emergence of sub-trajectories, and c) the expansion of the applications.

The identification of the initial applications, including the initial users and business models, is often full of surprises. Table 2.3 lists 13 industries created in the twentieth century where the initial users and business models were unexpected. Because they were unexpected, many firms did not initially focus on the appropriate customers or use appropriate the business models. This caused different firms, countries, and regions to have different levels of success in the new industries.

For example, while most people saw the radio as a wireless telegraph and thus believed that the main users for the radio were business users in the early twentieth century, a small group of U.S. firms found in the early 1920s that many people were interested in listening to music and other firms were interested in advertising their products on the radio programs. On the other hand, most European governments did not allow private individuals or companies to broadcast radio or television programs and they were also opposed to advertising. Furthermore, many European believed that U.S.-style radio and television programs were not appropriate for Europe's higher level of culture. Although it took many years for European firms to change their thinking, programs similar to those that succeeded in the U.S. market succeeded in the European market[3].

Two additional examples of unexpected customers and business models are the PC and mobile Internet.

TABLE 2.3. Cases of Unexpected Initial Users or Business Models

| Industry | Users/Application | | Unexpected |
	Expected	Actual (if different)	Business Models
Radio	Business Communication	Music for consumers	Advertising (not business fees)
Television	Entertainment for consumers		Advertising (not user fees)
Discrete Transistors	Military	Radios in Japan	
Transistor Radios	Unexpected market	Young people	
Mini-Computers	Unexpected market	Scientific and engineering	Sale and not lease
Open-loop NCs	Unexpected market	Low precision requirements	
Passive LCDs	TVs	Calculators, watches	
Internet	Unexpected market	Scientists and engineers exchanging mail and files	
Personal Computers	Unexpected market	Hackers, game players	
LANs	Mainframe computers	PCs (printing documents)	
Active matrix LCDs	TVs	Video cameras, fax machines	
World Wide Web	Unexpected market	Scientists and enginers creating and accessing home pages	Transactions and advertising
Mobile Internet	Business users	Young people	Micro-payments (not transactions)

FIGURE 2.1. Model of Industry Formation

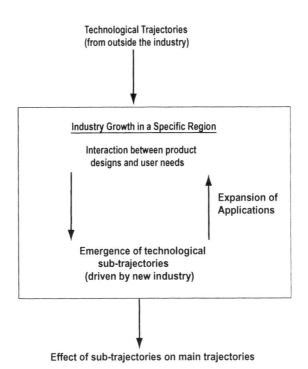

The PC Internet

The key technological trajectories that led to the formation of the Internet are terminals (dumb terminals, PCs, PDAs, or phones), network performance, software, and digital content[4]. Due to the constant improvements in electronics, which are often expressed in terms of Moore's Law, the processing speeds, memory sizes, and other characteristics of PCs, PDAs, and phones double every 18 months. Network performance also improves at a fairly fast rate partly due to advances in digital switching, the latter of which is also related to Moore's Law. The diffusion of computers has also caused the amount of software for managing networks and later digital content to increase in many industries.

These four trajectories led to the formation of the Internet first in the U.S. While the reasons for the early application of the PC Internet in the United States are complex and interdependent, one reason was the early identification of the first and still dominant application of the Internet, e-mail, in the 1970s. The early implementation of packet-based networks by the DOD, the early diffusion of computers in many U.S. organizations, and the experimental nature of U.S. universities are important reasons for the early implementation of e-mail in the United States in the 1970s and 1980s. And the network effects/positive feedback associated with e-mail strengthened the U.S. lead in e-mail usage and also led to the emergence of sub-trajectories in areas like routers, servers, and other aspects of packet-based networks, mail-based software, and other software for managing these networks also in the 1970s and 1980s.

These sub-trajectories led to improvements in the infrastructure needed to support the expansion of the Internet in the 1980s and also provided incentives for the development of the World Wide Web in the early 1990s. Key developments included the URL (Universal Resource Locator), HTTP (HyperText Transfer Protocol), HTML (HyperText Markup Language), and the browser in the early 1990s. These developments expanded the applications from e-mail to content access; this first occurred in universities and quickly expanded as governments allowed the Internet to be used for commercial applications. Again, while the reasons for the early use of these technologies in the United States are complex and interdependent, one reason was the early introduction of networked computers in US universities including both the Internet and Ethernet. Many of the first organizations to make their content available on the Internet were US universities and they created a critical mass of users that initially drove the diffusion of Web usage.

As is described below, universities have not played a role in the creation of a critical mass of users in the mobile Internet. This has caused mobile phone-related Internet standards to evolve in a much less "open" environment than PC-related Internet standards did. While Cisco, Sun and other providers of Internet infrastructure sold their equipment to universities who bought the equipment with government funds, mobile service providers must create their services in a much more uncertain environment. And firms will only attempt to develop these services if they perceive the chance for above-average profits, which will only be realized through services that are partially closed. Chapter 10 deals with this issue in more detail.

The Mobile Internet

The key trajectories that led to the formation of the mobile Internet are similar to those for the formation of the PC Internet except we substitute phones for PCs, wireless for fixed-line networks, and wireless network software for fixed-line network software. Improvements in electronics led to better phones and better networks, although the latter is also a function of new wireless technologies such as CDMA (Code Division Multiple Access). Similarly, different forms of software have emerged for wireless networks than for fixed-line networks.

A number of service providers introduced mobile Internet services in 1999, including NTT DoCoMo, KDDI, and J-Phone in Japan and many others in Asia, Europe, and the United States. Browsers in NTT DoCoMo's phones read compact HTML pages, those in J-Phone's service read similar pages (called MML), and most other browsers read pages that are based on the WAP protocol. The services in Japan and Korea are the first regions to find the initial applications that have led to the emergence of positive feedback and sub-trajectories.

As of mid-2003, NTT DoCoMo's i-mode services are the most successful services. When it introduced i-mode in February 1999, it also offered a micro-payment system, 67 firms offered contents on its official portal, and by May 1999 four manufacturers offered compatible phones. Banks made up the largest percentage of firms on the initial official menu (see Table 2.4). An appendix to this chapter provides some background on its implementation and the prices of its various services.

I-MODE: INITIAL APPLICATIONS AND USERS

The evolution of products and markets in the Japanese mobile Internet and actually in any industry reflect experiments by producers and users[5]. Initially a variety of product designs vie for customer acceptance, and some of them are successful. The success of a specific product with a specific customer causes other firms to copy the successful designs, and standardization of the product at some high level begins to occur. Furthermore, user needs also evolve as they learn how to use the new product. Thus, there is a simultaneous evolution in the product design and user needs.

In network products, a key part of this evolution in product designs and user needs is positive feedback and problem-solving. If a network

TABLE 2.4. Number of Firms on i-Mode Official Menu at the Start of Services

Type of Service	Number of Firms
Banking	21
Financial	12
Travel	6
News and Weather	6
Entertainment	5
Ticket Reservations	3
Train Information	2
FM Radio Stations	2
Recipes	2
Other	11

product achieves a critical mass of users, positive feedback emerges between the various aspects of the product (in this case the phones, contents, and services) and the users, and problem-solving occurs. Firms introduce phones with better mobile Internet capabilities, content providers introduce new contents, and service providers introduce new services because the number of users is increasing. During the first year of i-mode services, this positive feedback emerged and caused NTT DoCoMo and later J-Phone and KDDI to solve many of the same technical problems that existed but were not immediately solved in the United States and Europe. These problems included a lack of content, small displays, and high prices for phones and for J-Phone and KDDI long connection times.

Young People and Entertainment

The first key interaction between products and users in the Japanese mobile Internet was between young people and entertainment. The high mobility and ample free time of young people lead to their high use of the mobile Internet. Young people are major consumers of games and horoscopes, and their downloading of ringing tones and screen savers to "personalize" their phone is similar to their emphasis on personalizing clothing, hair, makeup, and handbags. Entertainment contents still represent the largest amount of traffic (see Table 2.5), and it appears that those young people who place a strong emphasis on personalizing their appearance are also big consumers of this entertainment.

The personalization of phones complements the emergence of phones as a form of status. Just as watches are considered jewelry more than a functional device, mobile phones and their related services bring status to their owners. And they are also conversation pieces. For example, while young men once brought their dogs to the beach to attract women, they now show off their mobile phones.

The early success of these entertainment contents in Japan caused

TABLE 2.5. Traffic on NTT DoCoMo's i-Mode Service
in 2002

Percent of packets transmitted by:	
Content accesses	85%
Mail	15%
Access percentage by content category	
Ringing Tones/Screen Savers	37%
Games/Horoscopes	20%
Entertainment information [1]	22%
Information [2]	12%
Database [3]	5%
Transaction [4]	4%

Source: NTT DoCoMo
[1] Sports, music, prize guides, TV, radio, magazines
[2] Weather, news (overlap with entertainment information),
   and lifestyle (cars, employment, apartments)
[3] Train and bus, traffic, maps
[4] Shopping, travel, financial

the number of entertainment content providers to increase dramatically
in the summer of 1999. Initially, few entertainment companies were
interested in i-mode, the general view being that it was "ridiculous to
have small pictures on a tiny cell-phone screen." However, in spite of
this general view, Bandai offered these screen savers in order to utilize
servers that were left over from a failed PC Internet service[6]. Bandai's
screen savers and Index's horoscope services were the most successful
services in the first six months of i-mode services; about 20% and 8%,
respectively, of the early i-mode subscribers subscribed to them.

The success of these contents caused the number of entertainment
content providers to rise from 9% of the total i-mode content providers
to more than 50% by September 1999. One of these new content providers was Giga Networks; it began offering the downloading of music
scores in the summer of 1999. Because the phones were not yet set up
to download the actual ringing tones, users copied the music scores
onto paper and then re-input them via the keypad. By September 1999,
Giga Networks had more than 100,000 paying subscribers.

The increase in the number of these entertainment content providers
naturally led to a rise in the number of i-mode subscribers, which caused
more entertainment-related firms to apply to become i-mode content
providers. These entertainment-related firms first applied to i-mode
and not the competitor's portals since i-mode offered a micro-payment
system and had more subscribers than KDDI. The increasing number
of subscribers and content providers enabled NTT DoCoMo to extend
the positive feedback to phones and solve many of the technical problems associated with phones.

This positive feedback was extended to phone manufacturers at the
end of 1999 when the second generation of i-mode phones was released. These phones included music synthesizing chips for handling
mini-versions of the MIDI (Musical Instrument Digital Interface) protocol in order to download higher-quality ringing tones. The two dominant mobile formats currently used in Japan are Compact MIDI and
SMAF (Synthetic Music Application Format). Files based on both formats are much smaller than MP3 or other recorded music files. The
MIDI protocol and music synthesizing chips that could handle greater
numbers of chords have made ringing tones the most popular type of
content.

Furthermore, because the most successful i-mode phone in 1999 had
a large display (a diagonal measurement of two inches) and it was clear
that screen savers, horoscopes, and games would look better on a color
display, other manufactures also introduced phones with large displays
including those with color at the end of 1999. Naturally the entertain-

ment content providers quickly upgraded their contents into color contents. The combination of better phones and contents caused the number of total subscribers to pass two million in December 1999, with almost 135,000 new subscribers being added each week.

User Learning

The second and third interactions between product designs and users reflect user learning. The second interaction was between the popularity of mail, entertainment contents, and phones with internal cameras. This interaction is one more example of how digital technology is reducing the production costs of contents if we define contents in a liberal manner[7]. Mail can be considered an expression of creativity and mobile mail has spawned an entire sub-culture including the use of "thumbs" and icons that serve as abbreviations[8]. Users like to add screen savers, ringing tones, URLs (including ones for their own home pages), photos, and beginning in early 2002, 5-second videos taken with an internal camera.

The small size of phone screens has also played a role in the high percentage of traffic that is represented by entertainment contents and mail. The small screens made it hard for users to access and utilize more complex business applications such as travel reservations, shopping, and mobile Intranets. The small screens also drive the importance of personalized mail services and to a lesser extent personalized site services in a wide variety of content categories like news (see Table 2.6). Both of these personalized services enable users to partly avoid the small keyboards and screens of mobile phones in their search for specific information on the site. And the addition of URLs to these mail messages (which is not possible with SMS) enables users to access more detailed information.

The third key interaction between products and user needs is between mail and portals. User learning drove a movement in traffic from official to unofficial sites in NTT DoCoMo's i-mode service. While the official sites are presented in menu form much like PC Internet portals, with unofficial sites the users physically input a URL or access the site through a saved bookmark or URL in a mail message, portal site, or search engine. The growth in traffic to unofficial sites was facilitated by the growth in mail since it is easy to include URLs in the mail messages. There was also a synergistic effect between the number of portals and search engines for unofficial sites and the traffic to the unofficial sites.

TABLE 2.6. Examples of Mail Services

Category	Mail Services
Entertainment	Daily screen savers
News and weather	Company or entertainer specific news, weather updates for specific areas
Retail	Sale information including discount coupons
Mobile shopping	
Concert tickets	Artist specific concerts
Music and books	Artist, author, or genre-specific music and books
Navigation services	Mail just prior to arrival at destination train station
Restaurants and bars	Information on new restaurants & bars (by type) including discount coupons
Travel (tours, airlines, hotels)	Mail on specific destinations or genre, airline reminders
Employment, cars, rentals	Mail on new jobs or rentals or newly available cars (by type)
On-line trading	Stock (for a specific firm) or index changes

User learning plays a role in each of the applications discussed in this book. While user learning has occurred first with young people in entertainment and mobile mail, user learning will continue to play a role in the expansion of applications for young people and in the expansion of the user base to business users. The former includes new applications like mobile marketing, mobile shopping, and the use of new technologies such as camera phones to photographs and access URLs, phones as mileage cards, tickets, and money, GPS-based location services. It is still unclear how the latter will occur; candidates include reading PC mail on a mobile phone, exchanging business card data through mail or an infrared or other connection between phones, and using mobile Intranet services.

Virtual Communities

The success of i-mode (and other mobile Internet and SMS services) has also caused a number of virtual communities to emerge and subsequent chapters describe some of them in more detail. As in the PC Internet, dating and chat sites are very popular in mobile Internet and SMS services. Services that facilitate the creation of individual sites are also popular since many young users like to include the URL for their mobile home page in their mobile mail. Some young people post their diaries and pictures of themselves, pets, and friends on their home pages; camera phones facilitate the loading of these pictures on sites.

The ease of taking these pictures has caused a number of potential problems to emerge in Japan and elsewhere. Young women complain about lecherous men holding cameras under their skirts, courtrooms and many firms have barred camera phones from their premises, shower rooms have become places to gather "exclusive" footage, and magazines are upset that some people are "stealing" their content by photographing magazine pictures. Some of these photos are shared in virtual communities while others are just used for old-fashioned espionage. Societies must come up with a variety of legal and technical solutions to these problems.

Returning to virtual communities, many firms are using this concept in their content services. Shiseido is trying to build a community of Shiseido customers. Net Price's shopping service facilitates communication between friends in order to obtain volume discounts. And many retailers and manufacturers are trying to create communities around their services and products.

Some of the most successive examples of content providers creating virtual communities involve the publication of mail magazines. Such services merely screen and choose mail magazine proposals that can be submitted by any would-be writer. Some of the most popular magazines involve diaries or commentaries on fashions. Virtual communities dedicated to various fashions have formed around these fashion-related mail magazines.

While most of these virtual communities involve legitimate exchanges, dating sites appear to also drive prostitution particularly with young women. Many women use the dating sites and camera functions to offer their services, and some of them will probably use GPS (Global Positioning Systems) in the future. On the other hand, many of these young prostitutes are quite willing to share their mail messages with the police when they are caught; professors and even judges have been arrested in Japan.

KDDI, J-PHONE, AND WESTERN SERVICE PROVIDERS

NTT DoCoMo's competitors, KDDI and J-Phone, began to become part of the positive feedback between phones, contents, and users in the Japanese mobile Internet in early 2000 as it became necessary for them to compete with NTT DoCoMo's i-mode service. While KDDI had introduced WAP in April 1999, it was the introduction of packet services in December 1999 and even more so micro-payment services in April 2000 that enabled growth in usage to finally occur. J-Phone introduced a micro-payment system and a quasi-packet system[9] when it introduced its mobile Internet services in December 1999. The micro-payment system was a prerequisite for offering entertainment contents, and i-mode's success made it easier to convince firms to offer contents on their portals.

The reason why KDDI and J-Phone solved these problems and the U.S. and European service providers initially did not solve them is because the success of i-mode provided evidence that investments in mobile Internet technology could pay off. The success of entertainment contents and the positive feedback that these successful contents created between contents, users, and phones provided this evidence. Evidence of a successful mobile Internet did not emerge in the European and U.S. markets in 1999 and 2000 because the service providers did not introduce micro-payment systems, which were a prerequisite for entertainment contents such as screen savers, horoscopes, and ringing tones. Instead they focused on business contents such as financial, travel, and shopping services that do not require micro-payment services. This prevented positive feedback from emerging between phones, contents, and users.

This situation began to change as the Western service providers analyzed the success of their short message services (SMS) in Europe and Asia. Although SMS was part of the early GSM standard, growth did not begin until the late 1990s, about the time that mobile phones had become popular with young people. More than 30 billion SMS messages were sent in December 2002, up from 2.5 billion in December 1999[10]. Although growth in SMS usage was initially slow in the United States due to incompatible systems between service providers, growth accelerated in late 2002 as the compatibility problems were solved[11].

The success of both i-mode in Japan and SMS in Europe caused the West to rethink the mobile Internet beginning with European service providers in the year 2000. Several European service providers have started i-mode services, Vodafone started Vodafone Live! in December 2002, and several US service providers started similar services in early 2003. For example, KPN, a Dutch service provider, started i-mode services in mid-2002 in the Netherlands, Belgium, and Germany. Bouygues Telecom started services in France in late 2002. Telefonica introduced i-mode services in Spain June 2003 and is the largest service provider in the world (44 million subscribers as of June 2003) to license i-mode from NTT DoCoMo. Italy's Wind is expected to start services in late 2003.

Vodafone's service has done somewhat better than i-mode in Europe. As of the end of July 2003, it had two million subscribers or more than three times the number of European i-mode services partly since it offered seven phones or five more than the European i-mode services did. While relatively old phones from NEC and Toshiba were only available for the i-mode services, phones from Nokia, Panasonic, Sagem, Sharp, Siemens, and Sony-Ericsson were available for the Vodafone Live! services. And the phones from Vodafone Live! included cameras and better displays, particularly the two phones from Sharp. The success of Sharp's and other Asian phones appears as if it will lead to an increased presence of them in the Vodafone Live! services at least in the short term.

Vodafone has done better than NTT DoCoMo because it has more effectively combined European and Japanese technology than NTT DoCoMo has. NTT DoCoMo attempted to transfer its i-mode system to Europe without making sufficient changes to the technical or supplier system. In particular, it insisted on the technical superiority of its c-HTML systems in spite of the success of WAP both in Japan and Korea. This lack of technical flexibility plus an emphasis on its historical phone suppliers has resulted in far fewer phones for the i-mode than Vodafone Live! services.

One downside of Vodafone's flexibility is greater variability in the standards that support its service. The service combines MMS and WAP; the latter includes a number of still undefined standards. These ambiguities have led to large differences in how handsets access contents, and these difference make it more difficult for content providers to write content for Vodafone Live than i-mode.

In any case, it appears that the introduction of these kinds of services, particularly the Vodafone Live! and i-mode services will change the global phone market. Mobile Internet services are evolving in a similar manner in most of the world in terms of contents and business models. Vodafone Live! and i-mode will likely continue to see growth in Europe and the rest of the world as Vodafone licenses its service to other service providers.

Furthermore, many of the services require phone manufacturers to design their phones to match the specifications that are set by the service providers. Up until now, service providers were expected to adapt their systems to the phones. The new strategy is partly aimed at Nokia, which has the strongest brand image in the market. Vodafone and other service providers would like to strengthen their brand images at the expense of Nokia's brand image.

The challenge for Nokia is to reassert its leadership in the mobile Internet, a theme we return to in Chapter 10 when I discuss platform leadership. Nokia is currently not doing this since it is the Japanese firms that are creating the solutions and applying them to Europe through their partners. Nokia is currently pushing the Open Mobile Alliance (OMA), which appears to be a rehash of the WAP Forum. These kinds of committees work well when the problem is easily specified, as it was in the days when air-interface standards were the main topic of consideration in the mobile phone industry. But the mobile Internet is a giant experiment between producers and users and committees work poorly in such an environment. Nokia will have to work hard to make OMAP competitive with Vodafone Live!, i-mode, and other services.

APPENDIX: THE CREATION OF i-MODE

Senior Vice President Keiichi Enoki is usually cited as the first person in NTT DoCoMo to recognize the possibility for a mobile phone-based consumer information service. He was looking for a way to utilize NTT DoCoMo's packet system, which was introduced for business users in 1997. Similar to other incumbents that have successfully introduced disruptive technologies, he created an independent organization and staffed it with outsiders. In mid-1997, he hired Mari Matsunaga from one of Japan's largest media companies, Recruit. She hired Takeishi Natsuno, an executive with an Internet company, and several other people, and together they created the concept for NTT DoCoMo's mobile Internet service, which is called i-mode[12].

Unlike in Europe, the success of mail and voice entertainment services with young people in pagers, PHS phones, and mobile phones beginning in 1997 further encouraged Matsunaga and Natsuno to focus on consumer services. In particular, the proprietary short messaging service offered by one of NTT DoCoMo's competitors (J-Phone), which was superior but incompatible to the other messaging services, including NTT DoCoMo's service, was a major hit with young people from the day of its introduction in October 1997. This caused J-Phone's share of young subscribers to skyrocket; in fact, it still had the highest market share among college students even as late as mid-2000, more than a year after the introduction of i-mode.

The success of this short messaging service provided Mari Matsunaga and Takeishi Natsuno with additional ammunition in their internal battle with those who preferred a focus on business users and a "rent-based portal." For example, an outside consulting firm strongly recommended that NTT DoCoMo charge firms for a position on the i-mode portal and let them be responsible for the collection of content fees. Mari Matsunaga and Takeishi Natsuno opposed these ideas and instead proposed the introduction of Internet mail services, a micro-payment system for firms on the official i-mode portal, and a method of accessing so-called unofficial sites with the input of a URL or bookmark. One reason for choosing Internet mail services was because they provided a higher level of inter-service compatibility than the proprietary J-Phone services did. NTT DoCoMo also adopted a dynamic default official menu where content sites are ranked according to their share of traffic, which changes somewhat from month to month.

NTT DoCoMo introduced i-mode in February 1999. Initially, it had 67 firms on its official portal, it offered four phones by May 1999, one of which had a rather large display (larger than two square inches), and it offered packet and micro-payment services (NTT DoCoMo takes 9% as a handling charge). The initial price of packets was 0.3 yen (0.25 cents) per packet, where each packet is 128 bytes or 1024 bits. It cost about 30 yen (25 cents) to access news or check banking balances, about 60 yen to send money electronically, and about 25 yen (20 cents) to make an online stock purchase. Although it initially cost 2.1 yen (1.7 cents) to receive and 4.2 Yen (3.5 cents) to send a mail message that has up to 250 Japanese characters (twice as many are possible for English characters) NTT DoCoMo and its competitors continue to drop these charges and as of May 2003, mail could be send for as little as 1 yen (.8 cents).

NOTES

[1] Source: GSM home page (http://www.gsmworld.com/news/statistics/index.shtml)
[2] For example, see Hargadon, A., *How Breakthroughs Happen: The Surprising Truth About How Companies Innovate*, Boston: Harvard Business School Press, 2003.
[3] For example, see Spar, D., *Ruling the waves*, NY: Harcourt, 2001.
[4] For example, data on the performance of digital devices can be found in (Woodal, P., "Untangling e-economics," *The Economist*, September 21, 2000) and data on the diffusion of digital devices can be found in (Lohr, S., "Technology climate is gloomy, but its future still seems bright," *New York Times*, July

29, 2002.
[5] This summary is based on a paper by Clark, K., "The interaction of design hierarchies and market concepts in technological evolution," *Research Policy*, 1985.
[6] Quoted in Koyama, T., "Bandai puts bad times behind it on road to success," *Nikkei Weekly*, April 16, 2001.
[7] Doyle, G., *Understanding Media Economics*, Sage Publications, 2002.
[8] Brooke, J., "Youth let their thumbs do the talking in Japan," *New York Times*, April 30, 2002, p. 14.
[9] J-Phone began charging by the packet before it introduced packet services in 2002.
[10] See the GSM home page: http://www.gsmworld.com/news/statistics/index.html
[11] More than one billion messages were sent in September 2002 up from 30,000 in June when compatibility became possible. See "Voice Still King of US Wireless Services," *Washington Post*, September 30, 2002.
[12] For more details, see Matsunaga, M., *The i-mode Story*, Singapore: Chuangyi, 2002.

Chapter 3

Key Technological Trends

Growth in the initial applications discussed in Chapter 2 has accelerated the pace and changed the direction of innovations in the Japanese and global mobile phone market. Prior to growth in mobile Internet services, the worldwide mobile phone industry benefited from better semiconductors and other improved electronic components, which were used to make better phones (primarily smaller and lighter phones) and networks. These better phones and networks, along with the digitalization of content and improved software, made the mobile Internet possible.

The success of mobile Internet services in Japan has caused a new set of technological trajectories, which I call sub-trajectories, to emerge from the main trajectories of phones, networks, contents, and software. These trajectories have emerged first in Japan due to the early success of the Japanese mobile Internet, and they are also driven by the higher subsidies that service providers pay retail outlets in Japan than in the United States and Europe. These high subsidies have enabled Japanese manufacturers to introduce much more technology (including mobile Internet technology) into phones than their Western counterparts and still produce a phone that within six months of its release is basically free to the final consumer.

Many of these technologies began appearing in Western phones in 2002, and many of the technologies were supplied by Japanese firms. Phones with polyphonic tones, color displays, and cameras were released in Europe and the United States in 2002, and we can expect continued improvements in the number of polyphonic tones, the color resolution of displays, and camera resolution over the next few years in these phones. However, these sub-trajectories have already begun to fizzle out in Japan as customer needs have been quickly satisfied, and we can expect similar results in the West in the next few years.

On the other hand, other sub-trajectories will probably represent key areas of competition in the mobile Internet in Japan and elsewhere for years to come and will cause more sophisticated applications to emerge; these applications are covered in Chapters 4 through 9. While most of these applications are very different from the main applications of the PC Internet, the sub-trajectories will also reduce and at some point in the far-off future probably eliminate the performance differences between the mobile and PC Internet. Improvements in display size, phone processing speed, memory size, and network speeds will lead to dramatic changes in the user interface. Improvements in the latter three parameters are already causing a movement from HTML-type pages to Java, and similar programming languages and further improvements in these three parameters will likely lead to completely new forms of user interfaces.

These improvements in the user interface and the expansion in applications are expected to accelerate growth in the Japanese market, which slowed considerably in 2002. More than 70% of Japan's mobile phone users owned a mobile Internet compatible phone by the end of 2001, and this passed the 80% mark in January 2003. And as shown in Figure 3.1, growth in usage per person has also slowed considerably as entertainment applications, which are still the most popular applications in the Japanese mobile Internet have become widely diffused.

As for suppliers of the technologies, early entry and the disruptive nature of the mobile Internet may provide some of the new entrants with strong first-mover advantages. The early growth in the Japanese mobile Internet means that many of these early entrants are Japanese firms or in some cases Western firms that are strong in the Japanese mobile Internet market. Of course some firms that are not strong in the Japanese mobile Internet market may be able to catch up due to their global strengths in the mobile phone (e.g., Nokia) or PC (e.g., Microsoft or Intel) markets. This will depend to some extent on the degree to which the mobile Internet places special demands on these technologies. For example, power consumption is much more important in the mobile phone than in the PC, and this dictates a different set of design tradeoffs in mobile phone than in PC applications. We first discuss ringing tones, color displays, and camera phones followed by display size, network and processing speeds, memory size, the user interface, and other network technologies.

POLYPHONIC TONES

The popularity of downloading ringing tones caused the number of chords or polyphonic tones that handsets are capable of generating to dramatically increase between 1999 and 2002. While the first phones could only handle a single chord, by the end of 2002 most phones contained chips from firms like Yamaha that could generate 40 chords. Nevertheless, it appears that few people can discern the difference between 32 and 40 ringing tones, and the difference between 40 and 64 is expected to become even more difficult to discern. Instead, the market is beginning to change to voice- and vocal-based ringing tones; and one of the drivers is network speed, which enables the use of short,

FIGURE 3.1. Packets Downloaded Per i-Mode User Per Day

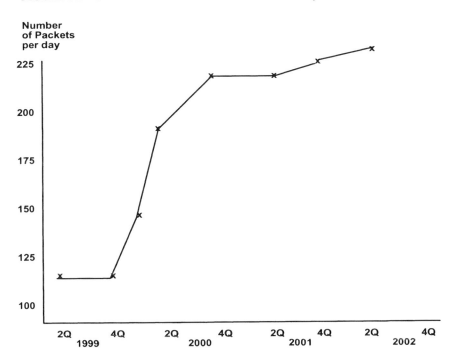

Note: 128 bytes per packet and 0.3 Yen or 0.25 cents per packet

CD-quality songs, as ringing tones.

COLOR DISPLAYS AND CAMERAS

The popularity of screen savers, games, and horoscopes has caused the quality of the mobile phone's color display to quickly approach that of PC displays. Phones with color displays were first introduced in late December 1999, and phones that could display more than 65,000 and 250,000 colors were introduced in the spring of 2001 and 2002, respectively. The first color displays were based on STN (supertwisted-nematic) technology while current displays are based on TFT (thin-film transistor) technology.

Furthermore, the availability of high-quality color displays, along with the popularity of mail and various contents, caused camera performance to become a key sub-trajectory. Japan's third largest service provider, J-Phone, introduced the first phone with an internal camera and complementary mail services (called Sha-mail) in early 2001. Pictures can be easily attached to mail just as with ringing tones and screen savers. Many types of content sites that support the exchange of photos taken with these cameras have also emerged. For example, it is possible to make still and moving screen savers from your photos, and the content services contain a variety of editing functions including the ability to add animated characters, borders, and music to the photos.

Although it is sometimes reported in the Western press that camera phones have much higher data usage (e.g., twice the amount) than regular phones, the actual increase is probably less than 5%. The reason is that few people actually send photos and after the initial novelty wears off, it appears that few people actually take pictures with their camera phones. The biggest impact of J-Phone's early introduction of the camera phone may have been on market share. Many industry observers argue that J-Phone was able to increase its market share by about 1% in Japan (from 16% to 17%) through its faster introduction of these services.

The number of pixels in the camera lens has been a key point of competition between phone manufacturers since their initial introduction. As shown in Figure 3.2, the cameras released in early 2003 had 1.3 million pixels, or almost enough to provide quality that is equivalent to the levels found in traditional photographs. As with digital cameras, the increased number of pixels in camera phones has caused competition to change from the number of pixels to other factors. For example, phone manufacturers have steadily increased the amount of digital zoom in phones, and some phones that were released in 2003 have digital zoom levels of 16 times.

A larger trend is the movie from photos to videos. J-Phone and KDDI introduced phones in 2002 that can take short videos (between 5 and 15 seconds), and the video can be attached to mail. Like the camera phones, it appears that few people send videos to other people. Nevertheless, the number of frames that are displayed per second and the length of the video are becoming key trajectories.

A key issue in both still and video camera phones is the relationship between the resolution, packet charges, memory, and processing speeds. Increasing the number of pixels, frames per second, and video length requires more memory, and more importantly the cost of sending such pictures and videos can become prohibitive. For example, a one million-pixel photo requires more than 100 kilobytes of memory and with NTT DoCoMo's service, can cost several dollars to send (see Figure 3.3). Of course, theoretically the service providers can offer consumers the choice of resolution when they take a picture and attach it to mail. Some people might want to send photos with a lower resolution than the ones they merely save in their phones.

Video resolution and the number of frames that can be displayed per

FIGURE 3.2. Trajectory of Digital Camera Resolution: Regular and Camera Phones

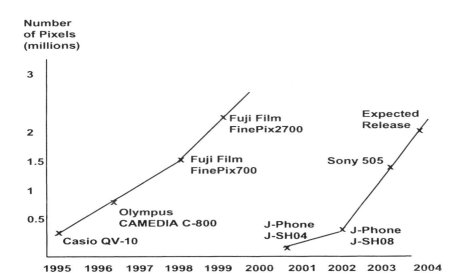

Note: 128 bytes per packet and 0.3 Yen or 0.25 cents per packet

second also depend on the speed of application processors. J-Phone initially introduced a custom version (with higher compression) of the MPEG4 multimedia file format due to the lower processing and network speeds of its phones. But since the phones it released in the summer of 2003 have much faster application processors, it has changed to the MPEG4 format. However, this still leaves the question of packet charges when you send a video file. J-Phone's custom version of the MPEG4 multimedia file format, which is called Nancy, could compress a five second low-resolution video file to 15 kilobytes, which is

FIGURE 3.3. Packet Charges for Sending a Single Photo

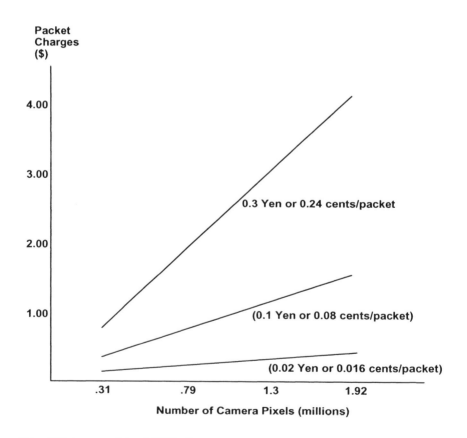

Note: 128 bytes per packet and 120 Yen/$

less than 10% the size of comparable MPEG4 files. At 0.3 Yen or 0.25 cents per packet, the price difference is very large: 28 cents versus $2.80. The evolution of processing speeds, memory sizes, and network speeds and costs will continue to influence the choice of compression techniques and other design decisions.

As an aside, many people believe that camera phones will cut heavily into digital camera sales over the next few years, and video camera sales will probably experience similar effects 5-10 years from now. The reason is that the camera lens and chip cost less than $25 in late 2002, and these costs continue to fall. Although processors and memory chips are also necessary and together can cost as much as twice this amount, the processors and memory are also used for other application such as Java, which is discussed later. Thus, the real cost of adding a camera to a phone is probably not much more than $25. This is why many industry observers expect camera phones to dominate daily photo-taking while digital cameras will be used primarily on vacations where zoom and wind angle functions are more important. But the advantages of camera phones will also increase as phone manufacturers increase both digital and optical zoom, the latter requires better lens technology.

The diffusion of camera phones will also likely accelerate the trend toward having fewer pictures developed, which was started by the diffusion of digital cameras. It appears that many people will take pictures with their camera phones and store them on their phone (or on external memory devices). Since most people tend to take their mobile phones wherever they go, they can show people their photos by merely displaying them on the screen. Larger displays will accelerate this trend. And like digital cameras, the diffusion of phones with external memory devices will support the transfer of photos to other devices such as PCs and televisions. By 2010, the largest and most profitable firms in the photograph industry may be those firms that provide image processing and photo management (e.g., albums) services.

DISPLAY SIZE

The change from a voice-only phone to an Internet-compatible phone has caused display size to become very important. Improved electronics had reduced the size of phones in Japan to less than 60 grams (2.1 ounces) by the end of 1998, which was seen by many as the physical limit for small phones. Smaller phones would not be able to cover the distance between the mouth and ear, and smaller keypads made it more difficult to make phone calls.

These improved electronics and their effect on the miniaturization of phones enabled manufacturers to apply the further improvements in electronics to the task of increasing display size just in time for the start of i-mode services in February 1999. The most successful phone among the first four i-mode phones released in the spring of 1999 had the largest display (which could display 100 Japanese Kanji characters), and it used a clamshell design to achieve this large display. The success of this 110-gram (3.9 ounces) phone enabled its manufacturer, NEC, to double its share of the Japanese phone market and replace Matsushita as the number one producer. Matsushita and other manufacturers have been forced to change to the clamshell designs, thus making a complete revision to their entire design strategy.

Since then NEC has increased the size of its displays from 2.0 inches (diagonal dimension) to 2.4 inches in its most recent phones. It is possible to further increase the display size using current technology at the expense of weight. Generally speaking, for a 110-gram phone, doubling the display size will double the display weight and increase the weight of the plastic housing by 50%. Similarly, a 20% increase in display area leads to a 30% increase in price. The issue is whether customers will choose a heavier and more expensive phone in order to have a larger display. Given the lower emphasis on weight and size in the European and U.S. markets, this is probably a larger possibility in those markets than in Japan.

Naturally the prices and weights of displays will drop in the future. As shown in Figure 3.4, prices have dropped and will continue to drop since improvements in yield and thus price reflect some of the same forces at work in the semiconductor industry. This will not eliminate the price disadvantage of larger display phones, but it will reduce the disadvantage.

The future of display weight is harder to predict. Unlike many of the other trends discussed here, display size does not depend on improvements in transistor density or manufacturing performance. Instead actual improvements in thickness and thus weight depend on the use of new display technologies. For example, displays based on EL (Electro Luminescence) produce their own light and thus do not require a separate light source. This increases the luminosity, the possible viewing angle, and the response time (better for viewing video), and it reduces both the thickness and potentially the power consumption.

Currently the most popular form of EL display is called organic light-emitting diodes (OLED). In 2003, Kodak released a digital camera that contains an OLED display. Sanyo Electric released a series of phones

FIGURE 3.4. Trends in Display Price

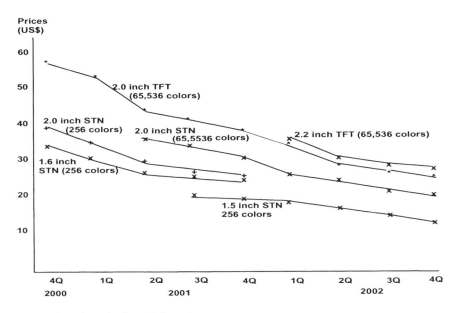

Note: Inches refer to the diagonal dimension.

with these displays[1] and it expected to sell one million of these displays in 2003. The display has 5 times the brightness, 2000 times the response speeds, similar levels of power consumption, and half the thickness of TFT displays. These displays are expected to become the standard by early 2004 as costs come down and durability (potential hours of usage) increases. Costs are primarily a function of yields, and both yields and durability are largely a function of experience. As of early 2003, displays only lasted for 10,000 hours or a little more than one year while most firms believe that at least 30,000 hours are needed.

Farther in the future are displays that are based on light-emitting polymers (LEP). Like EL displays, LEP displays create their own light and thus do not require a separate light source. The key difference is the use of an organic substrate. By applying a thin polymer film to a plastic substrate, firms can make displays that are thinner than one-tenth of an inch and can be rolled and folded. Thus it may be possible to double, triple, or even quadruple the size of existing phone displays. Color is also possible by spraying drops of red, blue, and green LEP material onto the substrate, and some U.S. manufacturers plan to release experimental color displays in 2003 or 2004.

A major problem is that silicon transistors, which are needed to switch the pixels on and off, are rigid, fragile, and operate at temperatures high enough to melt the plastic display. Thus firms are trying to develop organic thin-film transistors, which may take several years. In the meantime, these displays are mostly used in inexpensive applications like billboards and on T-shirts[2]. This makes LEPs a disruptive technology for mainline display manufacturers, and they may to lead to a new set of winners in this industry. One forecast made in 2003 predicts that these displays will be in mobile phones by 2007[3].

NETWORK SPEEDS

Network speeds are also important, as anyone who has sat and waited for contents to arrive either on their mobile phone or PC knows. For the second-generation phones that it released in 2002, NTT DoCoMo tripled its network speeds on to 29,800 bits per second. It also introduced third-generation phones and services (called FOMA) that are based on the W-CDMA standard in late 2001. These phones were initially capable of downloading data at 384,000 bits per second, and they are expected to have speeds of more than two million bits per second in the future. The growth in the number of third-generation subscribers has so far been slow primarily due to the cost of handsets and poor coverage but as these factors improve, it is expected that these phones will replace the second-generation phones.

KDDI's high-speed third-generation (3G) phones, which use the CDMA 2000 1x standard, have so far diffused much faster than NTT DoCoMo's 3G phones primarily due to lower cost and better coverage. At the end of August 2003, 9.7 million people owned one of KDDI's 3G phones while only 785,000 owned a FOMA phone. Unlike NTT DoCoMo's 3G phones, KDDI's 3G phones are not based on a completely different standard, and thus its new handsets are backwards compatible and have provided nationwide coverage from the start. Furthermore, KDDI has subsidized its third-generation phones much more than NTT DoCoMo has.

Nevertheless, KDDI still lags NTT DoCoMo in terms of data usage per mobile Internet subscriber, and its greater diffusion of third-generation phones has had very little effect on this difference. In the fourth quarter of 2002, KDDI's mobile Internet subscribers only generated 57% the level of data charges when compared to NTT DoCoMo's mobile Internet subscribers. And KDDI's faster diffusion of third-generation phones only increased its average data usage per subscriber slightly more than NTT DoCoMo between the second and fourth quarters of 2002 (16% versus 3.4%). Further, some providers of video services report similar levels of subscribers from both KDDI and NTT DoCoMo users in spite of the larger number of overall 3G subscribers for KDDI; this reflects the lower income of KDDI's subscribers and its weaker content base, which are probably interdependent.

KDDI's experience with 3G phones demonstrates the problems with assuming that network speeds are the only variable that is important in the mobile Internet. Content and, as we shall see in subsequent sections of this chapter, other innovations are also necessary to achieve strong and lasting growth in the mobile Internet. Some of these innovations are needed to handle the decompression of video, music, and other large data files that can be downloaded with the high-speed services while others are needed to improve the user interface.

In any case, these 3G services will play an important role in the mobile Internet in that the higher data speeds and lower packet charges will probably increase the usage of both new and existing applications. For example, due to frequency spectrum shortages in its second-generation system, NTT DoCoMo has not reduced basic packet charges since it started i-mode services in 1999, and the other service providers have only been slightly more aggressive than NTT DoCoMo. But most 3G services, including NTT DoCoMo's 3G services, use a different part of the frequency spectrum, and they use it more effectively than do the 2G services. This is why NTT DoCoMo and the other two Japanese service providers have set packet charges for their 3G services that are much lower than their 2G packet charges. Some are as low as 0.02 yen or 0.016 cents per packet, or 1/10 the level of NTT DoCoMo's 2G packet charges ($19.53 per megabyte of data at 120 yen per dollar).

Of course, the United States and European countries have much lower population densities than Japan and thus can more easily reduce packet charges in their 2G services than the Japanese firms. This suggests that once a critical mass of users and positive feedback between users, phones, and contents is created in Europe and the United States, growth could be stronger in Europe and the United States than in Japan. In particular, while i-mode packet usage per user stalled in 2002 three years after the start of services, it is possible that this will not occur in the United States and Europe due to lower packet charges.

In the long run, however, 3G services will provide much lower packet charges and thus will be successful. For example, Qualcomm estimates that the cost of sending one megabyte of data with 3G technologies is about 1/10 the level with 2G technologies such as GPRS. As shown in Table 3.1, Qualcomm believes that one megabyte of data can be transmitted for between 2.2 and 6.9 cents using 3G technologies; the lower estimate is about 1/1000 the current level of i-mode packet charges. If this occurs, it will be possible to download a three-minute song (using MP3), which is about 3 megabytes for a cost under $0.07 with cdma2000 1xEV. Similarly, it would be possible to deliver a two-minute medium resolution video clip (using MP4), which is about 6 megabytes, for a cost of about $0.13. Even if Qualcomm is off by a factor of 10, such reductions will probably lead to increased usage of the mobile Internet just as they have done in the PC Internet. With current packet usage stalled at between $10 and 20$ a month for the three service providers in Japan, such a stimulus is needed as was discussed at the beginning of this chapter.

FASTER MICROPROCESSORS

An unexpectedly important trajectory in the mobile Internet is the speeds of microprocessors since they are needed to decompress video, music, and other large data files and more importantly they can potentially improve the user interface without increases in the size of the display. In the PC world, large amounts of processing power existed before the Internet became widely used, and their prior existence has facilitated

TABLE 3.1. Cost to Deliver a Megabyte of
Data Traffic

Technology	Cost/MB
CDMA 1x	$0.059
CDMA 1xEV	$0.022
WCDMA	$0.069
GPRS	$0.415

Source: "The Economics of Mobile Wireless Data,"
available on Qualcomm's home page.

the diffusion of the Internet. For example, programs such as Microsoft Word, Adobe Acrobat, MP3, and MP4 reduce the data volumes - and thus reduce downloading times and costs - when exchanging text, music, video, and other files. Imagine having to exchange the Adobe Acrobat program each time you exchanged an Adobe Acrobat file.

Similar things will likely happen in the mobile Internet as processing power, and also memory and network speeds improve. In both the mobile phone and PC, improvements in processing speed and also memory size reflect Moore's Law. The shrinking dimensions of semiconductor chips have caused the speed of microprocessors and the size of memory to roughly double every 18 months for the last 40 years, and similar trends are already seen in mobile phones.

Of course the actual processors and memory that are used in phones are substantially different from their counterparts used in PCs due to different power consumption and power-up needs. Whereas power consumption in PCs and other non-portable devices can be greater due to those devices' greater heat-dissipation capabilities (and the obvious cord in the back), there are much larger concerns about power consumption in portable devices such as mobile phones. Manufacturers use lower voltages and other design changes to achieve lower power consumption in mobile devices.

These differences in design tradeoffs have caused the processors for PCs and mobile phones to follow completely different trajectories. Special-purpose processors, called base-band processors, have long been used in digital mobile phones to handle voice calls. UK-based ARM is the major supplier of these processors for GSM phones, while Intel is a major supplier for TDMA-based phones; Intel also acquired the largest independent supplier of them for Japanese phones.

The downloading of video and the use of Java (see below) has caused a new set of microprocessors, sometimes called application processors, to emerge. While initially video and Java programs were handled with the base-band processors, the expected growth in data-intensive applications has caused firms such as ARM, TI, Intel, and Hitachi to offer special-purpose solutions for these data applications. The fastest mobile application available in late 2002 had speeds of 133 MHz, and speeds of 500 MHz are expected by 2005 or 2006 with continued reductions in power consumption and price. By comparison, the fastest PCs had 2.2 GHz processors in late 2002.

At a more detailed level, the actual processing speeds will depend on the interplay between data processing needs, user behavior, battery technology, price, and the actual semiconductor processing technology. Strong demand for faster processors might lead to a steeper trajectory than we have seen with PCs, particularly when we consider that the motivation for higher processing speeds in PCs has come from greater complexity in Windows and its application programs, which many people consider unnecessary.

On the other hand, if mobile consumers actually use the applications made possible by these application processors, batteries will require more frequent recharging. The current distinction between talk time and standby time assumes little use of power when not talking. But as phones are used to play games and music, take videos, display GPS data, or exchange data with other devices, batteries can quickly require recharging even when the phone isn't being used for phone calls. Heavy users may be willing to purchase spare batteries as many PDA users do now or even use portable charging stations. For example, some Japanese trains (bullet trains) have plugs next to certain seats (mostly for laptops), and there are coin-operated machines that allow recharging of phones in some convenience stores (a typical fee is 100 yen or 80 cents for 10 minutes of charging time).

Third, power consumption concerns may cause phone and application processor manufacturers to use smaller semiconductor line widths to reduce power consumption and decrease processing power. Since they are linearly related, phone manufacturers could reduce power consumption to as much as one-fourth the current levels found in the latest 133-MHz processors while holding speeds constant through 2005.

Fourth, price determines the actual levels of processing speeds and battery times that can be offered in a phone. Phone manufacturers can offer phones with larger batteries (which also increases weight) and/or processors that use more advanced semiconductor technology. As of January 2003, the price of Hitachi's special data processor was about 5000 yen or $40, and some believed that the price of the future processors might fall to at least 1000 yen ($8) within a few years. For comparison purposes, the price of the microprocessors in current PCs is in the range of 10,000 to 20,000 yen ($80 to $160).

INCREASED MEMORY

Saving ringing tones, photographs, videos, and programs requires increases in memory, which will probably continue to occur through improvements in semiconductor technology and at the rate predicted by Moore's Law. Most phones released in early 2003 had more than 1 megabyte and several had more than 5 megabytes of internal memory. These phones could save as many as 2000 photos (taken with a 300,000 pixel camera), 2000 ringing tones (with 40 polyphonic tones), or 100 Java programs.

Some firms also offer external memory in much the same way that PC manufacturers did this in the form of 5.25- and 3.5-inch hard disks in the early 1980s. External memory is much cheaper than internal memory, and the most popular forms of external memory in Japan are Matsushita's SD memory card and Sony's memory stick. Mobile phones are the second largest application for this kind of memory following digital cameras; mobile phones use the smaller versions of this memory. Manufacturers expect to ship several million of the mobile phone versions in 2003.

As of early 2003, the 8-megabit version retails for less than 1000 yen ($8) while the 64-megabit version retails for 3000 yen ($25). But just as other forms of memory continue to experience cost reductions, the cost to produce external memory for mobile phones will continue to fall. By way of comparison, 8 megabits of memory can hold a 3-minute medium-resolution video or a little less than 10 minutes of music.

IMPROVED USER INTERFACE

Larger memory, faster processing capability, and greater network speeds are being used to improve the user interface, and it is possible that they can be used to significantly improve the user interface in the future through more client side processing of preloaded or downloaded programs. For example, fast application processors are needed for operating Macromedia's Flash, which became a standard item on phones in 2003, or for decompressing music and video files that are taken with a videophone or downloaded from the network or an external memory device. Initially, the more expensive and slower mobile network and mobile processor speeds caused J-Phone to use methods of compressing these files that are different from and of lower quality than MP4. Faster processors now enable J-Phone and the other service providers to use MP4 or similar versions of it.

Increased processing power can also potentially reduce the downloading time and costs of current text-based contents. Although this is not generally considered a problem with PCs, downloading speeds and costs are generally higher for the mobile Internet than for the PC Internet, and many content providers would like to move from text to graphical interfaces in order to compensate for the smaller screens and keyboards of phones. NTT DoCoMo's phones initially used browsers that interpret a markup language called c-HTML, which is a simplified version of HTM, partly due to the slower speeds, lower power, and smaller displays, keyboards, and memory capacity of mobile phones. Similar concerns drove J-Phone to use a markup language similar to c-HTML and KDDI and drove most Western firms to adopt WAP.

Java

The Japanese mobile Internet is currently undergoing a move from these markup languages to programs like Java and Brew, although it is possible to do similar things with other programs like XML[4]. NTT DoCoMo and J-Phone support different forms of Java, and KDDI (through its supplier Qualcomm) supports BREW. While there are differences between the two technologies, the key point is that users can download a Java or BREW program once and then utilize the program either independently of the network (e.g., with games) or in conjunction with data that are subsequently downloaded from the network. Alternatively, the Java or BREW program could be preloaded in the phone thus eliminating the need for downloading the program at all. For the rest of the book, I will refer to both BREW and Java programs as Java in order to simplify the discussion.

NTT DoCoMo introduced Java-compatible phones in January 2001, and games were initially the most popular content for these programs. Java reduces the cost of developing and downloading moving images that are written in HTML, and most games are now written in Java. Screen savers, horoscopes, and other entertainment contents that rely on moving images became large applications for Java programs in late 2002.

Many news and other text-based sites also began to offer their contents as part of a Java program in late 2002. Because most of the data involved in downloading HTML-based pages are tags that specify the format of the home page, Java programs that contain formatting information can reduce the amount of data to be downloaded and, equally importantly, reduce the time it takes to download the data. For example, a 30,000-byte program enables users to download up to six current c-HTML pages in one access, thus reducing waiting time and packet charges by as much as 80%. Many industry participants believe that most Japanese entertainment contents and most contents in general will be written in Java or a similar programming language by the end of 2003 and 2005, respectively.

The major limitation for Java and similar programming languages are the limited processing speeds, memory capabilities, and network speeds of current phones. Limited performance in these areas meant that the first Java phones (released in January 2001) could only download a Java program of up to 10 kilobytes and save between 5 and 30 of these full-size programs depending on the manufacturer. Since then, manufacturers have released phones that can download and process larger programs by increasing processing and network speeds and save a larger number of them by increasing memory. Larger program sizes enable more complex applications to be carried out.

For example, under ideal conditions it takes about 8 seconds to download a 30-kilobyte program with NTT DoCoMo's 504 phones that are compatible with its second-generation 29.8 kilobit/second service (there are 8 bits in a single byte). The cost is 170 yen ($1.42). Doubling the size of the program basically doubles the downloading time and costs, and doubling the network speeds basically reduces the time by 50%. Faster processing times reduce the time to activate a program, and lower power consumption enables the longer use of these programs. For example, the best current processor can activate a 40-kilobyte program in less than one second and allow the operation of such a program (e.g., a game) for up to two hours.

As of mid-2003, all three Japanese service providers had substantially increased the size of the programs that could be handled by their networks and phones. NTT DoCoMo and KDDI had set limits of 30 kilobytes and 80 kilobytes, respectively, and J-Phone had set a limit of 256 kilobytes that includes the total working space for the program. This working space is needed to run the program and handle the data that are utilized in the program. NTT DoCoMo offers a 100-kilobyte working space that is used to run the Java programs. Faster processors will expand the size of these programs and thus the complexity of the applications.

Limited memory also slows a move toward the use of Java programs even with some of the most recently released phones that can save more than 100 Java programs. For example, if the more than 3000 official content providers for i-mode all created their own Java programs from which users were expected to download information, users could only save a small fraction of these programs. Although increased internal and external memory will alleviate this problem, an alternative is for content providers to format their contents for standard Java programs. For example, it would be useful to have a single standard program for tables or for figures that each content provider could use to present their data on mobile phone screens.

Phones released in the spring of 2003 can access data for a Java program from different servers, thus facilitating the use of such standard Java programs. Added advantages of defining standard programs (or "objects"[5] for making these programs) would be less development costs for content providers and better control of viruses for service providers. Apparently it can cost more than 10 million yen ($80,000) to develop the largest Java programs, and concerns about viruses spreading through their networks has caused service providers to require extensive checking of Java programs before they can be downloaded via the service provider gateways. The emergence of standard programs would simplify these verifications, would enable smaller content providers to use Java programs, and will probably accelerate a move toward third-party development of Java programs. Of course, determining the appropriate standard Java programs and making your program one of these standard programs is a more complex issue. Several firms, notably K-Labs and Connect, are already striving to do this in the Japanese market.

Other User Interfaces

Increased processing power, memory, and network speeds may also improve the user interface in more radical ways. Some of these trends are evident in telephone call centers where voice recognition and synthesis are expected to replace most touch-tone menus over the next few years. These systems rely on natural language software that can be run on PCs. For example, most U.S. airlines allow callers to state the destination when they are requesting arrival information rather than choose a selection on the touch-tone keypad. And this is just the beginning. According to Forrester Research, 43% of North American companies either have purchased interactive voice response for their call centers or are conducting studies to do so[6].

It is just a matter of time before such interactive voice systems become available on phones. Japanese phones have had single-word recognition capability for many years and it appears that user resistance to making voice commands is a bigger bottleneck than actual technical problems. Speech synthesis has been available for years on Japan's car navigation systems. A phone released in 2003 by Fujitsu includes speech synthesis capability and it can change official contents and mail into audio; it is aimed at users with poor eyesight.

Another method is three-dimensional (3D) images and rendering techniques. While 3D movies and virtual reality require the use of special glasses, firms such as Sharp are trying to develop 3D displays for mobile phones that do not require the use of such glasses. Sharp introduced such a display in 2002, but a number of improvements are needed in both the display and peripheral circuitry such as processors and memory.

Three-dimensional imaging techniques, which have been available in some i-mode phones since late 2002, also enable information to be expressed in three dimensions. Processing speeds determine the quality of the images, and this quality is usually expressed in terms of number of polygons (any 2D figure with three or more sides) or polyhedrons (any 3D figure that contains polygons for each face) that can be represented. While phones with 50-MHz processors can display 10,000 polygons per second, phones with 500-MHz processors can display 640,000 polygons per second, which is currently available in the newest versions of the Playstation #2. The former is typically done at 1000 polygons per frame and 10 frames per second, while the latter is typically done at 10,000 polygons per frame and 64 frames per second.

A key question is how content providers can use such techniques to express greater amounts of data in, for example, 3D figures that can be rotated on the screen. While the initial applications for 3D rendering techniques are screen savers and games, as with Java it is possible to use these techniques in a wide range of applications. For example, 3D images of products might facilitate mobile shopping, 3D maps might facilitate navigation services, and 3D representations of data (which could provide the data on six faces of a cube) might facilitate business applications.

One way to make use of these 3D representations of data is to combine them with preloaded Java or other programs and either a more advanced version of Sony's Jog Dial function or a single word recognition system. Sony's Jog Dial enables users to quickly scroll up and down a page by rotating a tiny roller with your thumb rather than clicking down each line. An improved version of this function might enable users to also scroll in the horizontal direction in the same manner that we use the mouse to move in both the horizontal and vertical directions.

With such a function, users could choose one cube among several cubes that are displayed on the screen and rotate it in any direction, thus exposing a new set of data. If the data are downloaded into a program that contains all the tags needed to format the data on the cube face, it might be possible to represent more than 100 times the data than are currently shown on a c-HTML page, thus significantly reducing waiting time. Single word recognition and/or synthesis systems could also be used in place of the Jog Dial function to choose a cube, the cube face, and the specific information on each cube face.

More long-term approaches include virtual reality and holograms. With virtual reality it would be possible to walk through 3D worlds, pick up objects, examine them, and go to other Internet locations by flying or walking through doors. It is already possible to develop such 3D worlds using a computer language called Virtual Reality Modeling Language. The problem is that downloading and executing these files can both take more than several minutes even with the newest PCs[7]. Although the smaller mobile phone screens would reduce the size of these files, it could still be at least five years before useful applications appear.

Another futuristic approach involves the creation of 3D holograms that might be larger than the size of the display. A laser is split into two beams, one of which is reflected toward a piece of photo-reflective film. The other beam passes through an LCD screen that is connected to a fast processor. The beams overlap, forming a holographic pattern that is captured on the film and projected over the display[8]. This application is probably even farther off in the future than virtual reality.

OTHER NETWORK TECHNOLOGIES

New forms of networks will also play an important role in the mobile Internet. Firms are moving aggressively forward with Wi-Fi, one form of wireless local area network (WLAN). These networks may be faster and cheaper than 3G systems, and this is one reason why many Japanese and Western service providers are planning to make their phones compatible with Wi-Fi networks and enable their users to move between 3G and Wi-Fi networks. High processing speeds, like those found on Japan's latest phones, are needed to effectively utilize the Wi-Fi networks.

WiFi usage may also threaten an otherwise bright future for Qualcomm and other mobile infrastructure providers. In some ways, WiFi is to mobile infrastructure providers as PCs were to mainframe manufacturers. Whereas mobile infrastructure providers make their money through software and other custom functions, WiFi installations are increasingly being purchased from consumer electronics stores or on the Web.

Short-range communication technologies such as Bluetooth and various IrDA (Infrared Data Association) approved standards may also play important roles. Bluetooth has not done as well as expected due to its high cost and high power usage; both may be solved through better chips. KDDI released a phone containing Bluetooth that sold poorly due to its short battery life. Apparently the standby time was less than 24 hours; this is a major defect in a country where phones are expected to have 10 times this number.

Phones released by NTT DoCoMo in 2002 contain IrMC (IrDA Mobile Communications), a standard that was approved by the IrDA in the late 1990s, and IrDA Control, which is the protocol used in television remote controls. These phones are already being used in Japan to connect phones with cash registers, concert ticket machines, and each other (e.g., play games and exchange name cards) and control televisions and karaoke machines. It is possible that the former protocol can be used for two-way communication between phones and televisions, thus enabling phones to download some information from television programs. New infrared standards such as IrFM enable credit card information to be securely transferred between phones and cash registers. This protocol was established in 2002, and phones containing the protocol are being widely used in Korea for credit card verification; these phones will likely appear in the Japanese market in 2004.

Smart cards are credit-card-size devices that include an electronic chip that makes the card a small computer. GSM phones have contained contact-based smart cards for many years and credit card companies such as Visa and MasterCard have been trying to introduce these contact-based smart cards for years since they believe that the cards can reduce transaction costs and the rate of fraud. Recently, non-contact smart cards that rely on short-range (a couple of centimeters) radio transmission have diffused much more widely than contact-based cards. They are being used as transportation, concert tickets, and prepaid cards (<$50U.S.), and phones that contain these smart card functions are expected in 2004. When compared to smart cards, the required user involvement (pushing a button) of infrared techniques may increase transaction times, but this and the line-of-sight requirement may increase security. Furthermore, the line-of-sight restriction of infrared techniques and the short-range of non-smart cards may make both of these techniques more secure than Bluetooth.

On the other hand, Bluetooth may be more appropriate for applications that require longer transmission distances and lower security requirements. For example, Bluetooth could be used as an inexpensive W-LAN. More interestingly, Bluetooth may provide the basis for wearable computing. Many computer scientists have proposed the decomposition of the mobile phone into its basic components along with the attachment of them to various parts of your body. For example, it is possible to place batteries in your shoes that can be recharged as you walk and turn bracelets or watches into displays. Nike already offers shoes where the act of walking creates enough power to cause small lights on the shoes to blink on and off. And while current high-resolution color displays can be worn as bracelets or included in watches, improved displays will significantly improve the quality of such bracelets and displays. For example, displays that are based on light-emitting polymers can be rolled and folded, thus reducing the thickness of such bracelets. This would allow these bracelets to conform to the shape of your arm. Such developments could eliminate many of the performance differences between mobile phones and PDAs before the end of this decade.

SUMMARY

Several sub-trajectories will probably represent key areas of competition in the mobile Internet in Japan and elsewhere for years to come and will cause more sophisticated applications to emerge; these applications are covered in Chapters 4 through 9. Improvements in display size, phone processing speed, memory size, and network speeds will lead to dramatic changes in the user interface. Third-generation networks and W-LANs will also drive reductions on packet charges, and other network technologies such as infrared connection and smart cards will open up other applications.

NOTES

[1] "Coming soon to a laptop near you," *Economist*, June 19, 2003.
[2] "Move over, silicon," *Economist*, December 12, 2002.
[3] "Coming soon to a laptop near you," *Economist*, June 19, 2003.
[4] Microsoft is pushing XML.
[5] JavaScript is an object-oriented language, which means that it works by manipulating objects on a Web page.
[6] Roush, W., "Computers that speak your language," *Technology Review*, Vol. 106, No. 5, June 2003.
[7] For example, see Gralla, P., "How Virtual Reality Works," in *How the Internet Works*, Chapter 40, 2002.
[8] Freeman, D., "Are holograms finally for real? *Business 2.0*, July 2002 (http://www.business2.com/articles/mag/0%2C1640%2C41314%2C00.html).

Chapter 4

Phones as Portable Entertainment Players

Entertainment is still the leading content application in the Japanese mobile Internet and European SMS markets. The disruptive nature of the mobile Internet has enabled a new set of applications, mostly ringing tones and screen savers, along with a new set of users who are different from the previous lead users in the mobile phone industry to dominate the mobile Internet. More interestingly, it also explains why an unexpected set of firms have succeeded in ringing tones, games, and portals. Record companies initially ignored the ringing-tone market due to the poor quality of ringing tones as compared to music recorded on CDs. Instead, firms that sell karaoke-related hardware and software were the first firms to offer mobile Internet services (see Table 4.1), and they still largely dominate a market that is now one-tenth the size of the CD music market. Similarly, the largest video game firms initially ignored the mobile game market due to the poor quality of c-HTML games as compared to video games. They only entered as early as they did due to NTT DoCoMo's early support for Java, which has made it easier for them to offer simplified versions of their video games over the mobile Internet.

As long as the key trajectories do not change, the early leaders will likely continue to be the leaders in the Japanese market. Probably several hundred entertainment content providers make money, and the leaders are highly profitable. More than 10 firms have gross profit margins that exceed 30%. Since the official menus are primarily organized by traffic volume, the leaders have the most visible positions on these official menus, and these positions are a form of entry barrier. The leaders also use their high profit margins to spend heavily on advertising to create and defend brand image.

However, the trajectories are changing, and this will lead to further competition both within and between each type of entertainment contents. The key trajectories in the ringing tone market are changing from the number of polyphonic tones to network speeds and costs as improved sound quality begins to come more from the latter than the former. Although the size of the Java program will continue to be important in games and now in screen savers, it is the merging of ringing tones, screen savers, and games that will radically change the market for these contents. Screen savers can be turned into browsers and portable MTV-like services are already appearing. Furthermore, it is very likely that users will continue to demand services that enable them to create these own multimedia contents and share them with friends both online and in person.

Although it is far too early to forecast the winners, particularly outside of Japan, the differences between the technological trajectories in the mobile Internet and conventional entertainment industries suggest

TABLE 4.1. Leading Firms in Japanese Recorded Music, Karaoke, and Ringing Tones

Recorded Music	Karaoke Services (1999)	Karaoke Hardware (1999)	Ringing Tone Entry	Ringing Sales (2002)	Ringing Sales (2003)
Sony	Dai Ichi Fusho	Dai Ichi Fusho	Giga	Xing	Xing
Victor	BMB	Sega	Xing	Giga	Giga
Toshiba EMI	Taito	Yamaha	Dai Ichi Fusho	Dai Ichi Fusho	Dai Ichi Fusho
Pony Canyon	Sega	Roland	Sega	Sega	Dwango
Abex	Xing		Roland (NTT)	Yamaha	Sega
Universal	Taikan		Yamaha	Dwango	Yamaha
Papu			Bandai	Cybird	NEC
Warner			Taito	Roland	
Columbia			Namco	NEC	

Source: Digital Contents White Paper, Digital Contents Association, 2001, i-mode menu

that the eventual winners will probably have broad mobile entertainment experience. Conventional entertainment firms must develop mobile entertainment experience in order to develop the appropriate skills in Java and other technologies. Firms that attempt to merely use their existing content to compete in the mobile Internet will probably lose.

RINGING TONES

Music is big business all over the world and Japan is no exception. The market for pre-recorded music in Japan was about 500 billion yen ($4.2 billion) in 2001, down from a peak of 600 billion yen ($5 billion) in 1998. However, unlike other countries, karaoke also became a big business in Japan, much bigger than in the United States and Europe. In Japan, the popularity of karaoke spread to private systems in homes and so-called karaoke boxes in the late 1990s where it is not uncommon for groups as large as 50 people to hold a private party. Most of this karaoke music is downloaded via telephone lines as opposed to being played on special-purpose disks. The total market for these karaoke boxes and other services was greater than 1 trillion yen ($8.3 billion) in 1999, or twice as large as the pre-recorded music market.

New Entrants

It was the suppliers of karaoke hardware, services, and technology that first entered the ringing tone market in Japan. Giga Networks was the first firm to offer these services partly due to its relatively poor performance in karaoke hardware. Since the MIDI (Musical Instrument Digital Interface) technology[1] needed to play polyphonic ringing tones was not available in the first i-mode phones, Giga Networks started with services that required users to manually input the musical scores in the relevant place in the phone after downloading them onto the phone screen. In spite of the difficulties of copying the scores onto paper and then re-inputting them back into the phone, Giga Networks had acquired 100,000 paying subscribers by the end of September 1999, or more than 10% of the total i-mode subscribers.

Unfortunately, Giga Networks was slow to introduce ringing tones for the first phones that could use the MIDI technology. Instead, it was another karaoke firm, Xing, that first introduced such ringing tones. It did this through the help from Faith, which is a small spinout from the music firms Yamaha and Roland[2]. Xing had a monopoly with these services until Giga Networks and other karaoke firms like Sega, Yamaha, Roland, and Dai Ichi Fusho entered the market in February 2002. Initially the competition focused on providing ringing tones for the most popular songs, which Xing successfully did.

Furthermore, Xing and also the other entrants were able to use the presence of their hardware and software in the karaoke boxes to advertise their ringing tone services. For example, Xing has put advertisements for its ringing tone service in the karaoke catalogues that are in more than 200,000 karaoke boxes. It estimates that the customers of a single karaoke box look at these catalogues on the average of 10 times a day for a total of 2 million views a day. Beginning in 2003, it began sponsoring a music television program where it advertises its service and experiences a growing number of downloads just after new music is shown on the program.

The initial entry and success of these firms has provided them with a strong competitive advantage. As shown in Table 4.1 the order of market shares is quite similar to the order of entry, with Xing and Giga Networks still the top two firms; the major change is that Dai Ichi Kosho rose to number three through its strength in the karaoke market.

Late Entrants - Record Companies

The first large record company to enter the ringing-tone market was a consortium of record companies called Label Mobile that now includes five of the top six recording companies in Japan (Sony Music, Victor Entertainment, Avex Networks, Toshiba EMI, and Universal Music). Label Mobile was established in July 2001 and it started business in October 2001, almost 21 months after NTT DoComo started its i-mode services in February 2002. It began services for KDDI in December 2002 and J-Phone in March 2002.

These and other record companies were slow to enter the market due to the poor sound quality of ringing tones. They did not think people would pay for such poor music quality. But as the number of polyphonic tones has increased, the quality of the ringing tones has increased; furthermore, the record companies realized they could use ringing tones to promote the sales in their main market, CDs. Interestingly, the customers for CDs and ringing tones are similar, it was merely the difference in quality that caused the record companies to enter late. Of course, music companies outside of Japan (many of them own portions of the Japanese music companies) are well aware of the growth in the Japanese ringing-tone market and will most likely not make the same mistakes the Japanese record companies initially made.

In spite of their late entry, the Japanese record companies hope to use their information and technological advantages to overtake the karaoke firms in the ringing tone market. Their information advantages concerning new songs enable them to offer ringing tones for new songs at least two weeks faster than other firms. Furthermore, they also believe that they can integrate their promotion activities for CDs, concerts, and ringing tones, including the artist's official stamp of approval. The outcome is still unclear. Label Mobile's ranking had risen to number 16 by early 2002 and number 13 by mid-2002.

Oricon, which is the leading provider of music information in the Japanese market, also believes it can use its information advantages to compete in the ringing tones market. The Oricon chart appears in magazines, TV programs, and radio programs and it has used its contacts with artists and songwriters to offer ringing tones for new songs almost as fast as Label Mobile. And like Label Mobile it has created better ringing tones with a skilled artist who adjusts the ringing tones for each type of handset. Further, it is trying to promote its ringing tones in both its Oricon Club site, which offers information on 300 music clubs, and its screen saver services, which offer photos of popular singers and bands as screen savers.

A more interesting example is Dwango. As is described below, Dwango, who is not a supplier of video games, was the first firm to offer mobile games through its strength in the network technology used in video games. Its early success in providing mobile games also caused it to enter the ringing tone market. And in spite of starting its ringing tone services in June 2001 or about the same time as Label Mobile, its site has done even better than Label Mobile. A key reason for the success of the site is the ability of users to create their own arrangement of a specific song such as jazz, rock, or orchestra. This service reflects the desire of people to create their own contents.

Dwango's success is fairly consistent with the concept of disruptiveness. The mobile Internet offers a new combination of performance characteristics that appeals to a new set of users. In this case, it is the interactive nature of the mobile Internet that enables firms to provide an arrangement function in their services. A new entrant offers this new set of characteristics for a new set of customers. Demand for these services may grow just as the desire to create screen savers from photos taken with camera phones has grown (see below).

Basic Business Model

The ringing-tones market is very profitable, with more than 20 million subscribers paying between 100 yen and 300 yen per month in early 2003. And because their costs are so low, the profits represent more than 50% of the sales for the leaders. The major costs are ringing-tone production costs and, if you use someone else's song, copyright fees. The ringing-tone production costs are much higher for an original song, but they enable more differentiation.

The most common approach is to create ringing tones from existing songs. In this case the site must pay a copyright fee of between 5 and 7.7 yen (4.2 to 6.4 cents) to the relevant organization; in Japan it is called JASRAC. Thus, at the minimum, sites must charge more than 5 yen per download to cover these costs. Since users download an average of about three songs per month for the 100-yen per month services, the sites must pay JASRAC about 15 yen of the 100 yen in revenues.

These copyright fees had not been set in the United States as of early 2003, and the entire process for obtaining copyrights was still unclear as of early 2003. There are multiple institutions involved with these copyrights - for example BMI (Broadcasting Music Institution) and ASCAP (American Society of Composers, Authors and Publishers). The multiple institutions and the greater ambiguity in the process may slow growth in the United States. The U.S. owners of the copyrights may also push for higher fees since they have a much better understanding of the potential market of ringing tones ringing tones than when the fees were set in Japan. Other countries such as Germany, Singapore, and Korea also have multiple institutions and have more ambiguous processes than the United States.

Ringing-tone production costs range from 10,000 yen ($83) for basic ones that are made in China to 40,000 yen ($330) for ones made by a skilled artist who adjusts an "electronically" created ringing tone for each phone. One reason they are so expensive is that a different file must be made for each format and handset specification; handset differences include the synthesizer chips and speakers. Ringing-tone providers create a music melody file in MIDI format, transform it into a MIDI file using an authoring tool, and upload it to their Internet servers. The total production costs depend on the number of new ringing tones introduced each month; large ringing tone providers introduce as many as 300 new ringing tones per month.

We can use this data to create a simple equation for calculating the breakeven point for ringing tone providers. As mentioned above, the average revenues per subscriber are 85 yen/month and some content providers introduce as many as 300 new ringing tones each month. Thus, the breakeven point for the high-quality and low-quality strategy can be very roughly estimated using the following equation where RT refers to the number of new ringing tones introduced each month:

Revenues = Costs
85 yen X number of subscribers = 40,000 yen X 75 RTs;
breakeven point is 35,000 subscribers

85 yen X number of subscribers = 10,000 yen X 300 RTs;
breakeven point is 35,000

Although this is a very rough analysis, it is clear that firms with more than several million subscribers are making large profits (see Figure 4.1). Furthermore, it is possible that ringing tone production costs will be much lower in the West than in Japan due to greater emphasis on standardization - for example, in the MIDI format. This might provide even higher levels of profitability for the ringing-tone providers.

Voice ringing tones

Partly since it is becoming more difficult to improve the quality of ringing tones through increasing the number of polyphonic tones, ringing-tone providers are offering new forms of ringing tones such as voice and vocal ringing tones. Voice ringing tones play a recorded analog voice while vocal ringing tones play recorded music that includes the actual lyrics when there is an incoming call. The former can use the voice of virtually anyone including your own voice, animal sounds, or other sounds. Some of the firms that emphasize high-quality ringing tones claim that the voice ringing tones represent as much as 50% of their sales while most estimates are in the range of a few percent for all users and 15% for owners of the newest phones (as of late 2002).

There are technical, market, and business-model issues with both voice and vocal ringing tones. The technical problem for voice ringing tones involves the memory size. Just as more memory is needed to save ringing tones that include a larger number of polyphonic tones, more memory is needed to save these analog voice ringing tones. Most phones available in late 2002 only had enough memory for a five-second ringing tone, which raises the market question of who wants to listen to a five-second voice ringing tone repeat itself several times before you can answer the phone? Perhaps more relevant is whether

FIGURE 4.1. Estimated Profits of Ringing-tone Providers (annual basis)

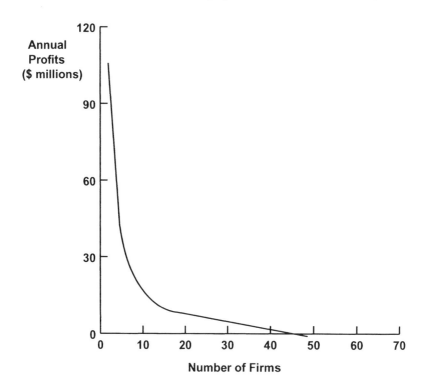

other people want to listen to your ringing tone repeat itself while they are trying to enjoy a meal in a restaurant. While larger memory sizes may reduce the repetitiveness of the voice ringing tone, these ringing tones may increase the number of people who would like to ban mobile phones from many public places.

These voice ringing tones may also require a new business model or at the minimum offer lower profitability. While voice-based ringing tones are not necessarily more expensive to develop than conventional ringing tones, it is likely that the license holders will charge more money than the one size fits all 5.0-7.7 yen per download. Industry people are now aware of the potential market for these ringing tones, and a celebrity is unlikely to let content providers receive the largest cut of this income as they do in current ringing tone services. Similar trends will probably also be seen in the United States where the relative success of the polyphonic ringing tones and later voice ringing tones will have a large impact on these license fees.

These technical, market, and business-model issues will likely lead to changes in shares in the Japanese and later other ringing tone markets. In particular, voice ringing tones involve new trajectories and thus require a new set of skills. Will the current leaders be able to use their skills in conventional ringing tones (e.g., development in China and experience with MIDI technology) or will they only have a brand image advantage that may be of limited value? For example, the leaders in the music market have stronger connections with voice copyright holders than do the current leaders like karaoke firms. It is possible that they will be able to obtain the rights to particular voices as part of a package and thus leverage their existing connections with celebrities.

Vocal Ringing Tones

Vocal ringing tones raise a different set of technical, market, and business model issues. Vocal ringing tones use a modified form of MP3 that requires less data but also has poorer sound quality than traditional MP3[3]. And even with the reduced data requirements, the vocal ringing tones require more memory and faster data speeds. This is why KDDI was the first firm to offer these services; it is the only service provider with enough subscribers to its high-speed data services to make such a music service worthwhile.

Label Mobile, in cooperation with about 15 music firms started offering 15 to 30-second samples of 300 songs in December 2002 where many of the songs include moving images and can be used as ringing tones. Users must pay 80-100 yen in content and about 170 yen in downloading charges (5-10 times the amount for conventional ringing tones). More than three million songs were downloaded in June 2003 and more than ten million had been downloaded since the service was started. This provided Label Mobile with almost 300 million yen ($1.5 million) in income and KDDI with almost double that in June 2003. The diffusion of 3G services and W-LAN will probably lead to reduced packet charges and thus lower fees for the vocal-ringing tones and music downloading services. If KDDI's vocal ringing tone service continues to see strong growth, it suggests that many people will be interested in downloading music to their mobile phones. This could reverse the pattern of lower growth rates each year for the ringing tone providers. While copyright fees paid to JASRAC rose 102% in 2000 and 72% in 2001, they only rose 30% in 2002 and it is likely that there will be very little growth in 2003.

Further, the diffusion of 3G services and W-LAN will probably lead to reduced packet charges and thus lower fees for the vocal ringing tones and music downloading services. If KDDI's vocal ringing tone service continues to see strong growth, it suggests that many people will be interested in downloading music to their mobile phones. This could reverse the pattern of lower growth rates each year for the ringing tone providers. While copyright fees paid to JASRAC rose 102% in 2000 and 72% in 2001, they only rose 30% in 2002, and it is likely that there will be very little growth in 2003.

If these vocal ringing tones do succeed, they may also shift the balance of power back toward the music companies since they own the rights to the music. The music companies who are participating in Label Mobile supply about 50% of the Japanese pop music; and like the United States and Europe, pop music represents a large percentage of the overall music market in Japan. On the other hand, the karaoke firms have the most visible positions on the official menus, and this may provide them with a negotiating advantage with the music companies.

GAMES

Video games are a major business worldwide. They started with arcade games, and now special-purpose consoles like the Playstation represent more than $U.S. 10 billion per year in sales. But like the ringing-tone market, the largest video game providers (see Table 4.2) were much slower to enter the Japanese mobile market than were the smaller firms. Only one of the three big suppliers of video games, Sega, had entered the market by the end of 2000. And of the next six largest firms, only Namco and Konami had entered the market by the end of 2000. The reason is that video games could not be easily transferred to the markup languages used in the mobile Internet. Only simple games can be written in markup languages such as WAP and c-HMTL, and even these simple games are expensive to develop since they require all character movements to be specified in the program.

Instead the early entrants were accidental entrants like Bandai and new entrants like Dwango, Index, and G-Mode. Bandai accidentally entered the market through the desire to use servers purchased from a failed service and a chance meeting with DoCoMo's i-mode group in late 1998. Dwango, a leader in network game technology (but not a video game provider), released one of the first popular games in late 1999 when there were less than two million i-mode subscribers. This game contained local data on real fishing places that included pictures

TABLE 4.2. Leading Firms in the Japanese Video and Mobile Game Markets

Video Game Sales	Mobile Game Entry	Mobile Game Sales (2002)	Mobile Game Sales (2003)
Nintendo	Bandai	G-Mode	G-Mode
Sony	Tomi	Hudson	Bandai
Sega	Namco	Bandai	Hudson
Namco	Dwango	Konami	Konami
Konami	Konami	Namco	Namco
Happy Net	Index	Sega	Sega
Square	Hudson	Enix	Enix
Kapukon	Sega	Dwango	Taito
Enikusu		Taito	Dwango

Source:s Digital Content s White Paper, Digital Content Associatition, 2002;
NTT DoCoMo's i-mode menu

of the places and fish. Users input the bait and time and subsequently receive information on their catch or lack thereof. It had obtained about 70,000 subscribers by early 2000, and the game was widely reported in the Western press (and sometimes used as evidence that i-mode would not succeed elsewhere).

Java

Java and other programming languages changed the mobile phone game market for both users and producers. For users, it reduced the need for connecting to the network each time they played a game. They could download a game once and then, depending on the payment scheme, play it for free. Some game providers only require users to pay a content fee when they download the game; others require users to pay a fee each month to play the game. In the latter case, they are automatically connected to the site (thus incurring some packet charges) each time they activate the Java program, and the site checks to see if they are a paying user.

Java also changed the mobile game market for producers since Java made it easier for video game providers to offer their existing games on the mobile phone. Java eliminates the need to program every action because it enables a set of simple rules to guide the sequence of events in the game. It also enables the movement of images and thus allowed the development of much more sophisticated games.

Even as early as 1999 the announcement of future Java compatible phones began to change the market because it encouraged the video game providers to enter the market on the expectations that they could develop mobile games that are based on their video games. Small video game software firms such as Tomi, Namco, Konami, and Hudson entered the market in early 2000. The similar customers for the mobile and video games also encouraged the small video game suppliers to enter the market.

These firms developed c-HTML games in order to build brand image on the mobile phone; and when the first Java phones first appeared in January 2001, they were ready with Java games. This helped the video game providers such as Konami and Hudson to become stronger partly at the expense of Dwango and Bandai. By the end of 2001, all of the new games were based on Java and most of the popular games were written in Java. The popularity of Java-based games has also enabled a new entrant, G-Mode, to become the leading provider of games in the Japanese mobile Internet.

Games also drove the initial sales of Java-compatible phones, the first ones being the 503 series from NTT DoCoMo. Game users were heavy users of pre-Java-compatible phones, and games were the first contents to use Java technology. This caused heavy game users to be the first buyers of the 503 series and caused the 503 users to have much higher packet usage than owners of the previous i-mode phones. Although NTT DoCoMo has continually emphasized that the higher packet usage by 503 users is evidence of continued growth in packet usage through new technologies, the fact that average packet usage for all i-mode users sees little growth suggests that this is not the case. Rather, the heavy packet users of pre-503 series phones were game users, and they were the first people to purchase the 503 phones, thus leading to the higher 503 usage.

Business Model

The profitability of games basically depends on obtaining sufficient subscriber revenues to cover development costs. The cost to develop games depends not only on the difference between html and java but also on the type of game. With Java, developers need to determine the game rules and then the programs can run automatically. Thus Java reduces the cost to develop a game but because it makes more complex games possible, Java games can cost the same or more to develop as html games. For example, one of the most popular types of games involves role-playing. These games have lots of scenarios, which cost money. For example, a popular Samurai game took many software engineers six months to develop. On the other hand, simple puzzle games have few scenarios and a few simple calculations.

Games are sold in packages or individually, and the service providers exert a lot of influence over the packaging of these games. For example, the most popular game category on i-mode is called game pack and as of early 2003 only two firms, Dwango and Konami, were allowed to offer these game packs. Dwango first offered 7 games in this package and as of late 2002 offered 14 somewhat complex games for 300 yen a month. Dwango and Konami's location at the top of the page provides them with a large advantage over the other firms.

The second most popular category of games also involves multiple games but the games are much more simple, often involving a single Java program. For example, Hudson offers 82 of these games in its mini-pack, and it is ranked number three in this i-mode category. Many of these games can be developed in a week or so and probably cost far less than one million yen to develop. Since it is introducing about 5 - 10 new games each month, it needs more than 10,000 subscribers to cover these expenses.

Industry observers estimate that the total market for game contents was 15 - 20 billion yen in 2002. Assuming 250 yen per subscriber per month, there are more than four million game subscribers. Industry observers also estimate that more than 200 game sites are profitable (as of late 2002). Of course, as with the ringing-tone providers, it is likely that the biggest firms are by far the most profitable. They achieve economies of scale in servers, other information infrastructure, and game development, and their high place on the menu provides a barrier to entry to other firms.

The Future of Mobile Games

Increased processing power and network speeds are leading to larger Java programs, more 3D images, and thus better games. While the mobile games will always be at a large disadvantage in terms of display size when compared to games played on special consoles like Playstation, it is likely that the processing speeds and thus the 3D images are only a few years behind the Playstation games. And as the 3D images became better for both phones and special game consoles, it is likely that the visual differences between the two will gradually disappear.

On the other hand, the increasing sizes of the Java programs and the use of 3D images will raise the development costs of these games. It currently costs about 170 yen to download a 30-kilobyte Java program and 300 yen a month for the rights to use the program each month in the case of NTT DoCoMo. Larger programs will raise the cost of downloading and the development costs. Doubling the size of the Java program will increase the downloading costs by about 25%.

For example, one magazine claimed that the most sophisticated games cost as much as 10 million yen or $80,000 to develop as of mid-2003. Breaking even on such a game would require more than 30,000 subscribers. Such higher breakeven points maycause game providers to offer fewer games since raising the price of games may be difficult to do. This is due to restrictions by some of the Japanese service providers and, more importantly, due to a possible unwillingness by users to pay more than 300 yen ($2.50) per month for a game. We will learn more about the willingness to pay as Western service providers set different prices.

Falling packet charges will reduce the cost of downloading games and also probably increase the usage of network games. Currently, such games are not very popular since the users can incur quite high packet charges. But as these packet charges fall, it is likely that a variety of multi-player games emerge. Treasure hunting and other such games that require many players might create temporary cults; the reduced packet charges might also cause GPS-based network games to become popular. Another possibility for network games is playing them offline using the infrared function. The requirement that the players be in the same place is a problem for games such as treasure hunting but perhaps not for others.

Higher processing and network speeds and reduced packet charges might cause mobile phones to become the dominant form of portable game. While it is unlikely that mobile games will challenge Playstation-type games in the foreseeable future, it is possible that they will replace other forms of portable games such as Game Boy. Even young children are likcly to use mobile phones to play games as the penetration rates for mobile phones continue to rise and parents start lending their phones to children to keep them quiet in the back seat on those long road trips. As with camera phones, portable games and mobile phones share a number of common technologies, and the advantage of mobile phones in terms of processing power will continue to increase.

SCREEN SAVERS

Other popular entertainment contents include screen savers and horoscopes. In these applications it is difficult to apply the concept of disruptive technologies since the performance and market for existing non-mobile phone-screen savers and horoscopes is difficult to define. How does one define the performance of a character or picture that is used as a screen saver or of a horoscope? While somebody provides horoscopes for newspapers and other media, what is the non-mobile market for screen savers?

Early market

In screen savers, it does appear that firms who own the rights to various characters have done well in the mobile Internet in spite of their late entry (see Table 4.3). Bandai, Disney, Universal Studios, Sanrio (Hello Kitty), NHK (Japan's public broadcasting station), and TV Tokyo all own the rights to various characters. On the other hand, firms who participated in other segments of the mobile Internet, including mobile phone manufacturers such as NEC and Panasonic or ventures such as Index and Cybird, were also quick to enter the market and either buy the rights to various characters or develop their own characters.

Interestingly, the rumor is that Disney was not successful in the Japanese mobile Internet until it enlisted the help of Cybird. In some ways, Disney should be the leading firm in the mobile Internet given its strength in characters, movies, and other forms of entertainment. Apparently, Disney's focus on licensing its characters as opposed to developing businesses around them has prevented them from developing a stronger position in the mobile Internet and will probably prevent it from becoming a leader in the future.

In the area of photos and art that are used as screen savers, firms that participated in other segments of the mobile Internet were also quick to enter the market. This includes Cybird, Index, Taito, and Hudson. But the early entrant and still leader is Gigno System Japan, which was established as Photo Net Japan in 1996 to provide digital image tech-

TABLE 4.3. Leading Firms in the Japanese Screen Saver Market

Number	Order of Entry	Mobile Sales (2002)	Mobile Sales (2003)
1	Bandai	Bandai	Bandai
2	Yoshimoto	Disney	Disney
3	Index	NEC	NEC
4	NEC	Universal Studios Japan	Universal Studios Japan
5	Atori	Sanrio	Sanrio
6	Disney	NHK	Yanase Takashi
7	Cybird	Cybird	Forside.com
8	Panasonic	TV Tokyo	Kodansha
9	TV Tokyo	Forside.com	TV Tokyo

Source: NTT DoCoMo's i-mode menu

nology in the PC Internet. It started providing i-mode services in June 1999 and subsequently entered into a number of licensing agreements with Ricoh, Plaza Create, Kodak Japan, and Sony. Many of these services were based on loading photos onto the Internet so that they could be accessed by a mobile phone. It changed its name to Gigno System Japan in December 2000.

The introduction of camera phones has made it easier to create screen savers from photos. While it was once necessary to visit a special photo shop to have your photos made into a screen saver, individuals can now do it themselves with their camera phones. Simply by using the menu for the camera function users can quickly create such a screen saver. Furthermore, various content providers allow consumers to customize the borders and add animated characters to these photo-based screen savers. It is also possible to assign photos to phone numbers so that, for example, the photo of the caller appears on the display.

New Technologies

New technologies such as Java, 3D images, and Flash are turning screen savers into multimedia entertainment. Java-based screen savers became possible with the phones released in 2002. These animation-based screen savers have menus that enable users to manipulate the images on the screen. For example, it is possible to walk the dog, make it roll over, and do other tricks on one screen saver offered by Panasonic. It is also possible to raise pets such as fish on other screen savers, an activity that was popularized in Bandai's Tamoguchi toys.

Larger Java programs enable more sophisticated manipulations and movements. The dogs walk and the fish swim faster, and at some point in time they begin to look like real fish just as it has become difficult to differentiate between animated and real characters in recent movies. They also enable more sophisticated menus. For example, the phones released in mid-2003 contain very small circles that serve as menu selections where users move the cursor between these five circles with the main control button on the screen. The small size of the circles makes them less obtrusive, causing them to detract less from the main characters on the screen.

More importantly, the interactive nature of Java programs along with larger programs enable users to download various data into the screen savers, thus turning them into more sophisticated forms of entertainment. The problem with this interactivity is the high packet charges that are incurred when downloading data. But just as network games will probably become popular as packet charges drop, it is likely that increasingly sophisticated screen savers will emerge that will become more and more like games.

Phones that can provide 3D images were first introduced in 2002 and phones compatible with Macromedia's Flash were introduced in 2003. Three-dimensional images provide a perception of depth, and Flash provides smoother moving images; these technologies increase the entertainment value of both screen savers and games and will probably contribute toward a merging of them. Java, 3D images, and Flash also raise the investment levels for entertainment content providers, which favors the large content providers. For example, the Macromedia authoring tool is expensive (50,000 yen) and requires skills that many amateurs don't have.

The phones released in mid-2003 also include the ability to link incoming calls and mail with Java programs; Bandai had announced such contents by May 2003. There are two methods of using Bandai's contents. First, if the sender and receiver are using the same program, the sender can create an application from the program and the application will be activated when mail or a call arrives at the receiver's phone. The other option is for the receiver to assign an application to a specific sender, thus causing the application to be activated when mail or a phone call comes from the specific sender. Like the moving images in KDDI's vocal ringing tones, these functions are another example of how the differences between ringing tones and screen savers will likely disappear.

Java will probably also change the market for those screen savers that are not based on animated characters. For example, Bandai planned to release a content service in the early fall of 2003 that enables users to combine photos into a video-like screen saver that includes music. The use of camera phones to create your own screen savers had threatened the screen saver providers including both the providers of animated and photo-based screen savers. But this may change as firms such as Bandai provide a service that enables users to personalize their screen savers in new and interesting ways. For example, a Bandai director showed me such a screen saver that included various pictures of his children flashing across the screen to the sound of his favorite song.

THE NEW VALUE CHAIN

The growing market and increasing technological sophistication of entertainment content has caused a complex value chain of activities to begin to emerge in entertainment contents and will likely spread to other contents. Of course, similar trends began long ago in the PC Internet but since the popular content and technologies are quite different in the mobile Internet than in the PC Internet, a new set of firms are carving out their niches in the mobile Internet.

Figure 4.2 summarizes the emerging value chain of activities. As discussed in Chapter 2, service providers collect content fees from users and pass on a percentage of these fees to the content providers. At the other end of the value chain, other firms offer content providers the raw content or services like content production and processing or site management. For example, Flex Firm is the leading provider of markup language transformation software, which enables contents written in c-HTML to be automatically translated into WAP. Be Map is a leading supplier of mail transformation software, which enables subscribers to different service providers to exchange mail. Photo Side.com is the leading provider of the ringing tone production services that were men-

FIGURE 4.2. The Emerging Division of Labor in Mobile Contents

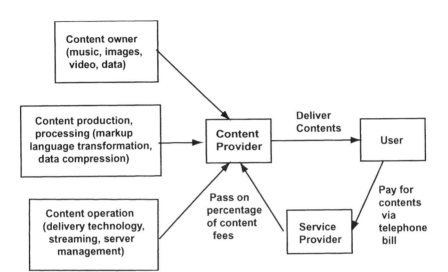

tioned earlier in this chapter. Kodak Japan is the leading provider of image-processing services for many of the screen savers mentioned above. Index operates more than 30 sites for other content providers.

However, the fastest growing market in the entertainment market is Java. Most new entertainment contents are already written in Java, and a similar situation is expected to emerge in non-entertainment contents. It appears that most entertainment contents and most contents Java requires a new set of skills that most content providers neither have nor appear to be developing. Other than game providers, the two leaders in this area are Cybird, Index, and their subsidiaries K-Labs and Connect; the latter two firms focus primarily on Java development for other content providers.

Although Cybird and Index have offered contents from the start of the i-mode services, they are not leaders in any one category (see Table 4.4). This is partly since they did not own any content in these areas. Instead, Cybird and Index have developed expertise in the underlying technologies through their early entry into contents and work with other

TABLE 4.4. Leaders in the Main Entertainment Contents (mid-2003)

Rank	Ringing Tones	Games	Screen Savers	Horoscopes
1	Xing	G-Mode	Bandai	Index
2	Giga	Bandai	Disney	Disney
3	Dai Ichi Fusho	Hudson	NEC	Telesys Network
4	Dwango	Konami	Universal Studios Japan	Cave
5	Sega	Namco	Sanrio	Cybird
6	Yamaha	Sega	Yanase Takashi	Work
7	NEC	Enix	Forside.com	Zappalas
8		Taito	Kodansha	Animo
9		Dwango	TV Tokyo	

Source: NTT DoCoMo's i-mode menu

content providers. As shown in Table 4.5, both firms are diversifying out of contents with Cybird's diversification primarily occurring through its K-Labs subsidiary. Skills in the underlying technologies such as Java will be important as the differences between the entertainment content categories blur and multimedia contents evolve.

INDEPENDENT PORTALS

NTT DoCoMo's decision to allow users to access any site with the input of a URL enabled the creation of an entire industry of independent sites. Just as in the PC Internet, there are hundreds of thousands of individuals who have created and probably millions who will create in the next few years their own home pages, and very few of these are intended to make money. As discussed in Chapter 2, these independent sites are an example of how digital technology is bringing down the production costs of contents. People use these sites to express themselves and communicate with others, and the growth in these sites is partly driven by the popularity of including your home page in mail

TABLE 4.5. Performance of Cybird, Index, and Their Subsidiaries
(April to August 2002)

Firm	Sales (Millions of yen)	Profits (Millions of yen)	Subscribers (millions)	Number of Contents	Percent of Sales from Contents
Cybird	4200	500	4.2	77	85%
Index	6400	600	5.8	>80	55%
K-Labs	830	Not released	NA	NA	NA
Connect	230	Not released	NA	NA	NA

Sources: Firm home pages. Sales of K-Labs and Connect (not publicly traded) assume 20 million yen per
year in sales for each employee.

messages.

Dating sites are one of the most popular types of independent sites in the Japanese mobile Internet largely because NTT DoCoMo does not allow them to be on the i-mode menu. On the other hand, they are one of the most popular categories on the official menus for the other service providers. Like their counterparts on the PC Internet, they facilitate communication between people and have created entire sub-cultures, some of which have contributed to prostitution and other social problems. Photos taken with camera phones are widely available on these sites.

There are also thousands of firms and probably tens of thousands in the future that will create sites that are either not intended to bring in income or at least do not require a presence on a service provider's official menu to make money. Just as in the PC Internet, it has become necessary for firms to provide information on the mobile Internet in order to reach those customers who are major users of the mobile Internet. Furthermore, unless a content provider needs access to a service provider's micro-payment system, it is not very necessary to be on a service provider's official menu. As the number of sites on the service providers' official menus increases, the value of being on the official menu decreases and content providers must find other ways to attract users to their sites, whether they are on the official menu or not.

The competition to provide portals and search engines for these independent sites offers another example of the disruptive nature of the mobile Internet. One would have expected Yahoo! Japan to be the leading independent portal since it is the leading portal in Japan's PC Internet. And if not Yahoo! Japan, one would have expected Digital Street, Giga Flops, or Softbank, all of whom were relatively early providers of search services in the mobile Internet. Instead, the greater importance of mail and young people to the mobile Internet than to the PC Internet has caused new firms to become the leading independent portals in Japan's mobile Internet.

Early Winners

One of the leading independent portal sites on the Japanese mobile Internet is called Girls Walker, which is operated by Xavel. Founded by in November 1999, Xavel did not start mobile services until April 2000, more than one year after NTT DoCoMo started i-mode services. However, while other portals and search engines focused on cataloguing sites, Girls Walker initially focused on mail magazines for women and used viral marketing to promote the creation of these mail magazines. Every mail magazine contains a link to the place in the Girls Walker site where people can propose new mail magazines.

Girls Walker was offering 17,000 types of mail magazines that are written by 1700 different writers as of mid-2001. The writers are responsible for writing mail and getting advertising income, although most of the mail does not contain advertisements since most people are doing it as a hobby. Xavel merely screens the mail magazine proposals and classifies them by genre and popularity. The most popular mail magazines are personal diaries.

Xavel makes money by sending targeted mail to its users and using this mail and its large number of page views to become one of the leading providers of shopping services. It had more than 5 million users in early 2003 (up from 3.4 million in mid-2001), and it was receiving 11.6 million page views a day (up from 8.4 million in mid-2001). Many of these page views are for the 75,000 sites registered on its contents portal as of mid-2001. The success of this portal and its mail magazines helps Xavel sell brand name clothing, accessories, and perfume for a number of manufacturers (see Chapter 6).

Magic Island, which is operated by TOS Corporation, has also utilized the importance of mail and the concept of viral marketing to become probably the largest independent portal in Japan's mobile Internet. Many young people like to include their mobile home pages in their mail messages, thus using their home pages as a communication tool. And each of the home pages that are created on the Magic Island site includes a link to the place on Magic Island's site where people can create their own sites. Its users had created more than two million home pages, and the Magic Island site and its home pages received more than 30 million page views per day in late 2002. TOS has made money by using Magic Island to lead users to its ringing tone site.

Although Girls Walker and Magic Island have not begun to use Java, clearly there are ways they can do this. Girls Walker could enable its mail magazines to activate applets and thus combine moving images with the mail magazines. Magic Island could offer services that enable users to make their own Java-based contents, including ones for their screen savers. One critical question is how these portals can obtain income from using new technologies such as Java.

Other Independent Portals

Other independent portals have not done as well as Girls Walker and Magic Island because in a narrow sense they have focused on contents and not mail. Broadly speaking, they have ignored the disruptive nature of the mobile Internet and instead have treated the mobile Internet like it is the same as the PC Internet. For example, Digital Street started its Oh! New? mobile portal in early 1999 at about the same time as NTT DoCoMo started its i-mode service. While Digital Street first focused on cataloguing sites, by August 1999 Oh! New? had achieved a sufficient level of fame that most unofficial sites were registering with Oh! New? By late 2000 it was receiving about 500,000 page views a day.

The problem with Oh! New? was that it focused on contents as opposed to mail and even for contents it was difficult to use. Digital Street did not screen contents until early 2001, and it did not provide rankings of the most visited sites. Instead, it provided a service that is similar to those found on the PC Internet where the problems of small screens and keyboards do not exist. It was not until Digital Street began providing a site creation service that it began to increase its page views, although its revenues were still a major question mark in late 2002.

Other sites have been somewhat more innovative and successful than Digital Street. For example, Softbank and Giga Flops assign four- to six-digit codes to sites for a small fee and provide a portal site where users can access these individual sites by inputting the codes as opposed to the URLs. Since the URLs for many of these unofficial sites can be as long as 30 digits, the four- to six-digit codes make it much easier for users to access the sites and then bookmark them for future reference. Softbank and Giga Flops do it in combination with magazines, which publish the codes and site rankings in terms of traffic.

Probably because of its success in the PC Internet Yahoo! Japan has relied the most of any Japanese firm on its PC Internet success in creating its mobile Internet services. It enables its registered PC users to customize their mobile Yahoo! portal in much the same way they can customize their PC portal. However, this is exactly how incumbents have failed in disruptive technologies. They target their existing users with the new technology and ignore the new and different users; in this case Yahoo! Japan has ignored the demand for mail magazines and home page creation services.

SUMMARY: MULTI-MEDIA AND THE FUTURE OF ENTERTAINMENT CONTENTS

The disruptive nature of the mobile Internet has caused an unexpected set of users, applications, and winners to emerge in the mobile Internet. Ringing tones, screens savers, and games are the killer applications and to some extent a new set of firms are providing these services.

More interestingly, mobile Internet-based entertainment will probably dramatically change over the next five years. Mobile phones may replace portable music players such as Apple's iPOD, which retails for several hundred dollars, since it is relatively easy to place the music playing capability in phones. One bottleneck is memory. As discussed in Chapter 3, 8 and 64 megabits of external memory retailed for less than 1000 yen ($8.30) and 3000 yen ($24.90), respectively, in early 2003 and the 8-megabit version can store almost 10 minutes of music. If these prices drop by 50% every 18 months, we can expect the 64-megabit version to reach $2 in about 5 years, with a 500-megabit version retailing for about $6.

Another issue is whether consumers will download the music via their mobile phone or PC or buy the music on pre-recorded external memory devices. This will depend on the relative prices and conveniences of the various services. As the prices of PC broadband and 3G packet charges drop, we might see a move away from purchases at retail outlets to PCs and mobile phones. Similar arguments can also be made for video, although the data volumes are much higher.

However, it is the unexpected changes that are of greater interest and I believe that the mobile Internet will create its own forms of entertainment just as the radio, TV, video recorder, and the Internet have done. Ringing tones and screen savers are merely the first examples of this phenomenon in the mobile Internet. For example, mobile phones may actually popularize so-called Web cameras since in some ways they can be more easily viewed from phones as PCs. Japanese service providers are hoping that daycare centers will install such cameras so that parents can check on their kids from their mobile phones. I believe that a more popular activity will be young people checking on the status of nightclubs as they look for Mr. or Miss Goodbar. Here, the mobile nature of bar hopping requires mobile phones while many worried parents have access to PCs in their offices.

The cost of such services will have a large influence on their success. Cameras that provide 360-degree views now cost about 70,000 yen ($583) and it costs about 70 yen to download a 15-second video using NTT DoCoMo's FOMA service (with a volume discount of 0.03 yen per packet). The video provides an aerial image of the location that is contained in a 300-kilobyte Java program. If the packet charges drop to $0.10 per megabyte of data, the packet charges for such a video would only be $0.03.

A second example is pornography, which has had major impacts on the start of the video recording industry and the Internet. Camera phones and video players make it easier for people to create their own pornography. Users are no longer dependent on the photo outlet for developing photos, and the small size of the phone makes it easy to take videos even when you are one of the participants. Whether or at least how these activities will translate into a business is another question. Currently, users attach photos and videos with their mail, and one reason why some service providers limit the number of times a single photo can be sent in a mail message is to restrict the sending of illicit and/or damaging material. But some entrepreneur will probably figure out how to get around this restriction and make money from people's desire to share their acting skills.

Another emerging trend in pornography is image-processing services. Image processing capabilities are reaching the levels where it is sometimes hard to determine the authenticity of a photo or video, and the impacts of this will be felt in the mobile Internet through improvements in the phone's application processor. For example, some Japanese firms have begun to offer image-processing services that replace bikinis with skin; users merely send the photo to a service and it comes back fully processes. In a world of Howard Stern, it's not hard to imagine processing services being offered that enable users to replace the participants in a porno video with other people by using pictures taken with the camera phone.

Merging of Existing Contents

We can make similar arguments about the future of existing contents such as ringing tones and screen savers. Technologies such as Java, Flash, and 3D images along with increased processing and network speeds and reduced packet charges will probably cause many of the existing distinctions between different entertainment contents to disappear. Label Mobile included moving images in its vocal ringing-tone services, and it is likely that these moving images will become more sophisticated. Although the early images were of snowboarders and other sport scenes, users will probably be able to choose from a variety of scenes in the future including animated scenes.

On the other hand, Bandai is adding music to its screen savers. Although it only discussed plans with me in June 2003 to add music to a single type of content service, this is probably only the first step. It is likely that the services that link Java programs with incoming mail and calls will also include music in the future. These trends suggest that the differences between ringing tones and screen savers will disappear. Furthermore, the Java-based 3D screen savers offered by Bandai will also likely begin to look like games, thus blurring the distinction between screen savers and games.

In this merging of the different contents, it is likely that personalization, interactivity, and virtual communities will play key roles. Users will probably continue to add their own photos and videos. The interactive capability will also enable users to personalize these "merged" contents by downloading data from the network and by exchanging data and these merged contents with each other.

An even more radical example of the merging of various forms of entertainment content is the use of this content as a browser. Some people argue that a 500-kilobyte Java program could be used as a browser. This would enable users to download contents or a second Java program from a Java-based program. The screen saver is an obvious starting point for such a Java-based browser. Content providers might first do this to encourage access to their contents as opposed to other contents while many Japanese retailers and manufacturers (see Chapter 5) are already thinking the same thing. If the so-called browser wars of the mid-1990s are any guide, there may be a major battle brewing for consumer eyeballs over the next few years in Japan and clearly elsewhere.

The use of screen savers or "merged" content as browsers may redefine the definition of a portal. If the screen saver used on a phone defines the portal, the value of the service provider's portal could be severely reduced. While this might cause service providers to restrict the use of screen savers as browsers, the potential benefits to users will cause somebody to allow it, and then there will be pressure for all service providers to offer similar services.

A more fundamental problem with using screen savers or other merged content as browsers may be the battery power needed to continuously run a Java virtual machine, which is necessary in order to quickly activate a Java program during an incoming call. A 500-kilobyte Java program will require a fast processor to activate the program, and this processor may use a lot of power. Of course it is only a matter of time before such processors are available. And this timing is both a function of the issues discussed in Chapter 2 and the number of ways that consumers will want to use their phones. Playing music, games, and videos consume even more power than does using a Java program as a browser.

In any case, even without the use of screen savers and browsers, increases in processing speeds, memory capability, and network speeds will make it possible for firms to offer more sophisticated entertainment that begins to compete with conventional entertainment. Downloading music on phones as opposed to buying it in CDs is probably merely the tip of the iceberg. Nevertheless, mobile entertainment will likely remain distinctive from traditional entertainment through the ability of users to personalize the entertainment for their own phone.

The Winners

It is far too easy to predict the winners either elsewhere or in Japan. The similarities between these multimedia screen savers and MTV suggests that major entertainment firms could be the winners. However, the mobile Internet is very disruptive for MTV, Disney, and similar firms since it is currently difficult to offer services with quality that is equivalent to that on MTV. Thus, many Japanese incumbents ignore the mobile Internet in Japan, and Western incumbents are likely to initially do the same or at least not focus on the right services. Firms that attempt to merely use their existing content to compete will lose, particularly those that think they already have the "premium" content or perfect business model needed to compete in the market.

It is more likely that the winners will have developed a broad set of skills in mobile entertainment as Index, Cybird, and Bandai have done. These skills will include the use of Java and other client-side technology that are optimized for the low processing and memory capability of phones. The rumor is that these firms already have a great deal of influence over the specifications for Java in phones. And as the processing speeds, memory capability, and network speeds of phones increase, these kinds of firms can expand these skills and the markets they serve.

On the other hand, firms who focus on a limited form of entertainment content like the leading ringing-tone providers may be in danger. They need to move toward vocal ringing tones and think of how they can integrate other contents with the vocal ringing-tone services. As for firms who are strong in conventional entertainment, they can also win, but they must develop these new skills in markets that are currently far smaller than their main markets. This issue is part of the larger issue of platform competition that we return to in Chapter 10.

NOTES

[1] NTT DoCoMo adopted a simplified MIDI format for its i-mode service, J-Phone adopted SMAF (Synthetic Music Application Format), and KDDI uses both. Faith developed the simplified MIDI format and Yamaha developed SMAF.

[2] Faith's chance came through a fortuitous interaction with a firm called Giga Rex (no relation to Giga Networks). Giga Rex proposed a simple form of MIDI technology for ringing tones to Mitsubishi Electric's phone manufacturing division. Mitsubishi asked Faith for advice and with Giga Rex the three firms approached NTT DoCoMo. For a variety of reasons Faith provided NTT DoCoMo with one of its employees for the development of the relevant mini-MIDI standard. Simultaneously, Faith approached Xing and offered its technology in return for a 50% share of the new business. Faith had profit margins that exceeded 80% in 2002 while Giga Rex has been far less successful.

[3] KDDI's service uses a smaller sampling rate than that used on the PC Internet and on CDs.

Chapter 5

Mobile Marketing

Mobile phones have already become the new contact point with young people in Japan through the success of the mobile Internet and to some extent in Europe and Asia through the success of short message services (SMS). Japanese consumers, particularly young consumers, are already using their phones to access and redeem coupons, request and receive free samples, respond to surveys, and access product information. For example, more than 100,000 Japanese redeem coupons with their mobile phones each month in Japan's leading video retailer, Tsutaya Online, and the total redeemed in Japan could easily exceed one million a month.

Firms provide these services on the PC and now mobile Internet because they are cheaper, faster, and more effective than traditional methods. Sending discount coupons via electronic mail costs far less than placing them in newspapers. It is faster and cheaper for firms to receive and respond to requests for free samples through electronic mail than via postcards. Conducting surveys on the Internet is also cheaper and faster than in phone interviews.

And the mobile Internet mail offers some very large advantages over the PC Internet mail with only some disadvantages. It enables firms to reach a different set of customers than can be reached with the PC Internet. The greater reach of the mobile phone means faster response times, the ability to carry the information into retail outlets, and greater opportunities for synergies between the mobile Internet services and existing information systems. The only disadvantage of mobile versus

PC Internet mail is a slightly higher cost to phone users; they must pay a couple of yen to receive or send a mail message with their mobile phone (sending mail from a PC to a mobile phone is basically free for the sender).

The focus on young people represents the disruptive nature of the mobile Internet. Young people are the main users of the mobile Internet, and the trendiest young people are the biggest users. Many of these trendy young people spend more time watching television, hanging out in malls, and reading fashion magazines than reading newspapers and books, and the mobile Internet complements their lifestyle much more than the PC Internet does. Japanese firms that sell to young people have moved the fastest to make the mobile phone their new contact point for customers. These include video rental outlets such as Tsutaya Online, beverage manufacturers such as Kirin Beverages, and cosmetics manufacturers such as Shiseido. In fact, Tsutaya Online's parent Culture Convenience Club had record profits in 2001 largely due to the success of Tsutaya Online's mobile Internet services and how they increased the number of video rental customers.

New technologies and their effective integration with existing information systems will continue to expand the opportunities for retailers and manufacturers. Reading bar codes with POS scanners or linking phones and registers with infrared functions enables retailers to integrate mobile information with POS information and use phones as mileage/point cards. More than twenty firms now offer such systems in Japan. Small devices that activate a phone's mail function facilitate the updating of mail addresses and thus sending mail to these members. The mail can also be used to activate a Java program that will enable richer information to be exchanged. The Java program might be provided to consumers in the form of a screen saver that also provides a form of advertising for the retailer or manufacturer. Location-based services enable consumers to register for location-based discount coupon services.

These technological improvements will also expand the number of users to include not just young people but people of all ages. For example, if business people begin to use phones for exchanging business cards or downloading information to printers in airports, it is possible that user learning will make the mobile phone a new contact point for them. And the involvement of business users in these activities will increase the level of disruptiveness for traditional suppliers of customer relationship management software and services. Many of the key mobile Internet technologies that are being developed for retailers and manufacturers are coming from a new set of suppliers that have found niches from where they hope to expand their market. Incumbents need to understand the key technological trajectories and the evolution of user needs.

This chapter looks first at the implications of the mobile Internet for discount coupons followed by the distribution of free samples, conducting surveys, building brand image, attracting mobile members, linking databases, and using phones as mileage/point cards.

DISCOUNT COUPONS

Discount coupons are used as a promotional tool for new products and to attract price-sensitive shoppers. Price-sensitive shoppers take the trouble to look for discount coupons, while most of the rest of the rest of us pay the higher prices. The mobile and PC Internet is a faster and less expensive way to distribute discounts coupons than traditional methods, and as we shall see later, a way to improve the effectiveness of discount coupons through the integration of these services with existing information systems.

The advantage of the mobile Internet is especially large with young people who tend to read newspapers much less than their parents. In most cases the mail can be sent within one day with the latest price and other information while newspapers sometimes have a two-week lag. And while the PC Internet has similar advantages over newspapers, the mobile Internet has the further advantage that phones can be carried into stores, restaurants, and other places of business. Printing out discount coupons is troublesome, and people are much more likely to remember their phone than a piece of paper.

Tsutaya Online: Video Rentals

Tsutaya is the largest video rental chain in Japan and Tsutaya Online is its fully owned online subsidiary. It was the first retail outlet to recognize the importance of the mobile Internet in Japan and in the world. It began offering mobile and PC Internet services in August 1999, and it was an early experimenter with these services. It was very successful in convincing its customers to register for mail services, and it began sending discount coupons to the mobile and PC mail addresses in the form of bar codes in the spring of 2000. Users merely showed these bar codes, as they are displayed on the mobile screen or to a lesser extent versions printed out from their PC, to clerks at the register to obtain a

50% discount on video rentals. The clerks visually confirmed the discount coupons, and initially the stores did not use scanners to read the bar code or record information about the specific customer.

These discount coupons have had a dramatic effect on the number of customers and rental income. For example, Tsutaya Online first sent the discount coupons to online members in November 2000. It found that during the time period in which these mobile coupons were valid, the number of store visits by online members increased by 45% as opposed to only 9% for regular members when compared to the week before. Furthermore, the sales from these online members increased by 41% for online and 10% for regular members again when compared to the week before. The similar increase in sales and store visits (in spite of the use of discount coupons) suggests that the people who used these discount coupons purchased more items than regular customers.

Since then, Tsutaya Online has continued to expand the number of people registered for its mobile Internet services along with the variety of its services. A major part of this effort is to convince card members to become online members since Tsutaya can communicate with online members much cheaper through mobile and PC mail than through postal mail. The number of its online members passed the 3 million mark in late 2002, which represented almost 20% of its total members, and continues to grow.

These PC and mobile mail services continue to be the main form of communication between Tsutaya Online and its members. Between 80% and 90% of its members register for one of the more than 100 types of PC and mobile mail that are offered by Tsutaya Online. On the average, users register for 3 - 4 types of mail; mobile mail has more than twice the response and click rates than PC mail. Although almost all of the mail is free, it contains entertainment, product, and campaign information, as well as advertisements. As discussed in Chapter 6, it has used these mail services to become one of the leading mobile shopping provider in Japan, with about $US 20 million in 2002. Furthermore, unlike the rates for PC and mobile banner advertisements, the rates for mail advertisements have not declined much for firms with strong brand images such as Tsutaya Online, and its income from these mail advertisements exceeded $US 10 million in 2002.

Jeansmate: Clothing

Jeansmate is the second largest clothing retailer in Japan. It started providing discount coupons on the mobile Internet in the spring of

2001 through NTT DoCoMo's special portal (see below). Users followed the link to the Jeansmate site, where many of them registered to become online members and receive mail that contained discount coupons. And many of these online members, more than 60% of them, were not Jeansmate members at that time.

Like the case of Tsutaya Online, customers merely showed the bar codes as they were displayed on the mobile phone screen to store clerks. These coupons provided a maximum discount of 20%, and they were provided in 500-, 1500-, and 3000-yen increments. Although each of them represented about one-third of the number of coupons redeemed, the 3000 yen coupons represented more than 50% of the value of those coupons redeemed.

Jeansmate only send these discount coupons every few months in order to balance the costs of discount coupons with the benefits of attracting new customers. Jeansmate wants to attract new customers who will then become repeat visitors once they have become members and receive the benefits of membership such as the point system. The point system creates a certain level of switching costs just as mileage programs or other mileage programs do. Jeansmate also tries to time these campaigns with fall, winter, and summer clothing sales.

The effect of these discount coupons has for the most part increased with each campaign. As shown in Table 5.1, there were more sales, customers, and sales per customer in the more recent campaigns as compared to the first campaign.

TABLE 5.1. Increased Effect of Subsequent Mobile Campaigns by Jeansmate

Campaign	Date	Sales	Number of Customers	Sales per Customer
First	April 2001	100%	100%	100%
Second	August 2001	695%	681%	102%
Third	November 2001	1068%	745%	143%
Fourth	January 2002	825%	631%	131%
Fifth	April 2002	1866%	1080%	173%

First Kitchen: Fast Food

First Kitchen is the second largest fast food provider in Japan, a distant second after McDonald's. It began an online membership program in early 2002 to augment its card-based membership system. As with the card-based system, members receive a 10% discount on all sales. In the online system, users enter a URL or send mail to an address that is advertised in the fast food outlets. More than 6000 people became members within two months of First Kitchen starting the online membership systems as compared to an overall total of 130,000 members.

First Kitchen created an online member system in order to reduce the costs of its traditional card-based membership system. It expects the online membership system to cost two-thirds less than the regular system, primarily by eliminating the need to send cards through the mail. However, it has created a two-tier system where the mobile users can get coupons, but they cannot acquire points. Only card-carrying members can acquire points for further discounts partly for technical reasons (see below) and also to maintain a sense of exclusiveness with the card-carrying members. The card-based membership system was created with exclusiveness in mind; it advertises the membership system very infrequently.

Location-Based Technologies

Many observers have proposed using GPS to send people location-based discount coupons and other advertisements. These commentaries often describe a world in which people are bombarded by discount coupons and other advertisements from the stores and restaurants in their area. For example, the movie *Minority Report* showed a world in which billboards speak to people as if they are one of the persons' best friends and people's movements are monitored by eye-scanning equipment.

The problem with these descriptions is that most people don't want to be bombarded with information and have their location's constantly monitored. This has certainly been the case with both PC and mobile mail where individuals, companies, and governments have fought against Spam. People don't like Spam, and thus location-based discount coupons must be services in which the user registers for them.

Of course it's still not clear how many people will register for mail to receive location-based discount coupons and other advertisements. For example, if you were planning to meet a friend in a city's downtown area, location-based discount coupons would only become useful when you reached the downtown area. By this time you would probably have an inbox full of mail messages that tell you about restaurants and stores in the areas you have just passed through. Being able to activate the service after you have reached the downtown area might be useful (but probably difficult without dramatic changes in technology). Furthermore, schedulers that include route guidance services and automatic updates using real-time weather and traffic information might enable location-based discount coupons to become more useful. Chapter 7 deals with these types of schedulers in more detail.

A Japanese railway company is just implemented a simpler approach. Holders of its train passes can register their mobile address and their interests. When they pass through a ticket wicket, they receive one or more mails that correspond to the interests they have registered. The mail might include discount coupons for local restaurants or information about sales in a local shop.

FREE SAMPLES AND GIVEAWAYS

The mobile Internet is also changing the way in which free samples are distributed and the type of free products that are distributed. The number of firms that distribute free products through their PC and mobile Internet home pages continues to increase. For example, after a cosmetics firm sent mail to 45,000 people who had registered for such services, 9000 people opened the mail and 1000 people requested the free products that were offered in the mail.

The number of giveaways that are related to mobile Internet services is also increasing. Many firms offer free ringing tones, screen savers, and games to customers instead of coffee cups and other products that contain the firm's logo. For example, Nestle's Japan offered free ringing tones for the song that was played on TV commercials for its Nescafe Cappuccino campaign. In the first week, 120,000 of these ringing tones were downloaded, thus giving Nescafe additional exposure each time one of these phones rang. And the cost of sending these ringing tones was almost nothing compared to the cost of the TV commercial.

Firms are now starting to give away free Java programs and if packet charges significantly decline, it is likely that free music and video clips will also become popular. For example, Cybird offers an ASP service for Java-based screen savers where it customizes several standard Java programs for firms. These screen savers change according to a preset schedule so that firms can provide their customers with marketing updates.

Kirin Beverages is one of Japan's largest beverage companies and it was the first firm to recognize the effect of the mobile Internet on the distribution of free products. It began offering a free Internet game to customers that purchased its Kirin Fire coffee drink in March 2001. The users accessed the game by sending a 12-digit number to a mail address that was revealed when a seal is pealed off the can.

The game is a simple high-low game. Users must guess whether the face down card is higher or lower than the face-up card by clicking on high or low. If they guess correctly three times in a row, they become eligible to win prizes like a lighter or a watch. Winners input their address, name, and other information to become eligible for such prizes.

The game was a huge success. The games were played about 10 million times during the initial campaign in a 10-week time-period in the spring of 2001. On the first day, there were 140,000 accesses, and by the end there were 240,000 accesses per day. Many people played the game more than 100 times.

More than one-half of the accesses were from a mobile phone. The number of mobile accesses was slightly more than the number of PC accesses in the morning and daytime but after 5 P.M. the PC became the dominant method of access. Surveys found that many people played the game right after they bought the drink from a vending machine, which is where most Kirin Fire drinks are sold. Kirin places the URL at the top of the vending machine.

Kirin managed to increase sales, particularly with new customers and obtain mobile mail addresses. Kirin experienced a 10% rise in sales during the campaign period partly through the acquisition of new customers, who hopefully will remain customers. While most coffee drinkers are older and tend to be blue collar, the mobile campaign attracted young computer users. The additional sales and profits were far more than the cost of game development and putting seals (less than 1 cent) on the cans.

Kirin also accumulated a large number of mobile mail addresses, which it plans to use in various advertising campaigns and mileage programs. Kirin advertises its sponsored events with mobile mail. It plans to enable users to acquire points from buying drinks (recorded as they play a game) that can be used to obtain prizes. It might use the mail to temporarily discount products that aren't selling well, particularly in combination with advertisements on other media. For example, it could send mail to its members in the days following a major TV campaign or include the URL next to magazine advertisements.

SURVEYS

Surveys are also more inexpensive and faster to carry out on the mobile Internet than on other media. Surveys on the mobile Internet can be carried out in less than a few day and the costs are often less than a few hundred yen per mail message with response rates sometimes higher than 50%. It is not uncommon to receive 30% response rates within three hours of sending the mail. This type of fast response time also enables surveys to more effectively ask about time-of-day activities something that is hard to do with other media.

For example, Research International conducted an analysis of tea drinking habits in Tokyo in October 2000[1]. It wanted to know when and where people buy and drink different types of nonalcoholic beverages. The questionnaire asked respondents about their last two drinking occasions. By spreading out the times in which the survey invitation was sent to potential respondents, Research International was able to obtain responses from different times of the day.

There are other advantages to the mobile Internet. It is easy to reward respondents with phone bill credits. Phone commands can be embedded in the survey to allow the respondent to register verbal responses to open-ended questions via a voice call to a telephone server. Each mobile phone has a unique ID number that can be verified when respondents access the survey, which provides an additional method of ascertaining respondent identity.

One obvious disadvantage to the mobile Internet is the small screens on mobile phones. Thus brevity is absolutely critical including scrapping the use of long brand names and minimizing the effect of scrolling bias. Scrolling bias is common in PC mail surveys, and firms try to minimize the effect by randomizing the order of selections. In mobile Internet surveys, it is best to place questions on separate pages so that respondents must scroll through the entire page before proceeding to the next question. Other considerations in mobile Internet surveys include ways for respondents to resume their responses after an interruption, which is common in most wireless situations.

New technologies like Java and video will continue to change this market. Java provides a better user interface including the increased ability to use graphs and other non-text information. One current disadvantage of Java is greater difficulties in ascertaining phone ID and thus information about the respondent. As with free samples, significantly lower packet charges would make it possible to include video clips with surveys.

An interesting variation on mobile-Internet based surveys is the use of lotteries as an incentive for survey respondents. While many sites offer prizes and phone credits to the survey respondents, a firm called EV Net offers respondents the chance to win between 10,000 and 10 million yen by responding to surveys. Respondents first register their age, gender, job, and income on either a mobile PC site. At the end of August 2001, it had more than 200,000 members: 75,000 were mobile and 125,000 were PC users. The purpose of EV Net's surveys is to measure a respondent's understanding of advertisements for clients. Respondents look at advertisements, answer questions about the advertisements, and receive a 13-digit lottery number for correct answers and win money when their lottery number is chosen. Clients determine the questions and pay EV Net fees for having these surveys completed. Clients include theme parks, banks, wedding centers and horoscope sites.

BRAND IMAGE

The mobile Internet is also a tool that can be used by some firms to enhance their brand image, particularly those that are targeting fashion conscious young people. These fashion-conscious people may own a PC, but they spend too much time at parties, malls, and other public gatherings to be at home. And when they are at home, they're watching television or talking on their phone. The mobile Internet, in combination with other advertising media such as TV and fashion magazines, helps firms reach these fashion-conscious young people; a previous section discussed examples for Kirin Beverages and Nestle. Mobile mail is currently the key technology, but Java and even screen savers will play a role in the future.

A leader in this area is Shiseido. Shiseido established PC and mobile Internet sites to enhance its brand image, not to sell cosmetics. Brand image is critical in this business as manufacturers are careful in how and where they sell their products. Shiseido sells some products in department stores, some in drug stores, but none in both. The highest priced products are advertised in magazines and not on TV. Shiseido's PC services are more popular with mature users than with young people. This is why information on how to stay young has the most traffic on its PC site while information on beauty has the most traffic on the mobile site.

Shiseido divides the cosmetics market into three categories. At the bottom are people with some interest in Shiseido's products, in the middle are customers, and at the top are big fans. Shiseido has devised different mobile mail services for moving people from the bottom to the middle and the middle to the top category. The former mobile mail services include information on seminars and how to receive free products. As of late 2001, Shiseido was receiving about 70,000 requests per week for free products and it was using a lottery to choose the 20 to 5000 winners.

The latter mail services include information about nails and a "build your own profile" service. Through a set of multiple-choice questions, the latter service asks women to describe physical characteristics and views towards beauty. Shiseido uses this information to assign the women to one of 45 types of mail services; the mail includes advice on beauty and information on Shiseido's products. As of the fall of 2001, 20,000 people had built their own profile: 70% of them were in there twenties, 20% were in their teens, and 10% were greater than 30.

ATTRACTING MOBILE MEMBERS

Sending mail to mobile members is one issue; another issue is how to attract mobile members. Spam doesn't work very well and it can have a deleterious effect on a new industry. Therefore, much more sophisticated methods are needed to attract members. Firms that attract these members can strengthen their relationship with their customers and/or help other firms do this. Service providers and firms with successful portals are doing the latter.

Firms that help other firms send discount coupons and free samples, conduct surveys, and build brand image are one part of the mobile marketing communications market in Japan, and both of them overlap with the mobile advertising market. In Japan, the mobile advertising market reached an estimated 5 billion yen in 2002, or twice the size of the market in 2001 but still a small fraction of the total advertising market in Japan of 6 trillion yen. Most of these mobile advertisements are mobile mail services rather than banner advertisements, and they may include discount coupons, offers of free samples, and surveys.

Successful Advertising Portals

NTT DoCoMo quickly recognized the power of its successful mobile Internet service and began working with Dentsu, Japan's largest advertising firm to offer these kinds of services. They created D2C in 1999 and had 300 million yen in sales in 2002, with 100% growth rates expected to continue well into 2003. Their site has received more than two million page views per week since early 2001 and like other mobile Internet sites, the main peak is at 10 P.M. and a second peak is at noon. Most of the clients have little to do with the mobile Internet and include financial (loan services for young people), food and beverage, cosmetics, publishing, and entertainment firms, roughly in that order.

KDDI and J-Phone basically copied NTT DoCoMo; their portals, done in collaboration with other advertising firms, have also seen rapid growth. KDDI's advertising portal expects to have one billion yen in sales in the year ending March 30, 2003 or more than twice the level of NTT DoCoMo. Successful content providers such as Tsutaya Online are also creating popular advertising portals. Tsutaya has used the success of its discount coupon services to offer similar services for manufacturers. People can pick up free ramen noodles in a Tsutaya Online store, and they may also rent a movie or buy a CD while they are there. Firms with both successful bricks and mortar and online sites can offer such services

Do It Yourself

Building your own site, particularly one with a large number of users is not easy, as many firms have found in the PC Internet. This is why many Japanese firms use the advertising portals that were created by the three services providers to reach these young people. In the short run, it is cheaper and easier to do this. However, in the long run firms need to create successful mobile portals if they are to use the phone as the new contact point for customers. This is particularly essential to those firms who are targeting fashion conscious young people.

There are several ways to do this. First, firms can use the advertising portals that were created by the three services providers to initially attract members to their site where they can offer discount coupons and other information. Jeansmate did this. Second, retail outlets and restaurants can sign up customers and do this in combination with traditional mileage/point card systems (dealt with later). Third, manufacturers can offer free ringing tones, games, or other mobile Internet-related services such as Kirin Beverages has done to attract members.

Fourth, they can use other media such as TV, magazines, and posters to attract users to their site. Of course this method was widely used in the U.S. PC Internet and is now used by many Japanese firms to sign up mobile members. Here the large reach of mobile phones and new technology is making this much easier to do with mobile phones than

PCs.

Cybird, a leader in mobile entertainment content, developed a mail service called Sugu mail, which literally means immediate mail. Since most people find it easier to send mail than input URLs, particularly mail that has a short address like the Sugu mail address, this service compensates for the difficulties of inputting long URLs on the small mobile keyboard. Users merely send mail to a Sugu mail address and quickly receive a return mail with the relevant information or a URL, which they can click on to obtain more detailed information. For example, Tower Records puts such a mail address on posters in train stations and on its store registers. Simply by sending mail to the address, users can receive mail about events like concerts. In an even more dramatic example, such a mail address printed on posters for a popular artist's concert attracted 1.8 million responses in three weeks.

By putting a 3-digit number in the mail, users can request more detailed information. For example, Keihama, a medium-sized train company, has assigned numbers to each station; by putting the number in the mail the user receives a timetable for the specific station. One of Japan's largest cable TV companies, WOWOW, uses these numbers to send information about specific programs to potential customers including details about the actresses and actors. It puts such a mail address in free newspapers that are distributed in electronic stores. In Kanebo's (a large cosmetics firm) service, users put their height and weight in the mail and receive mail about the appropriate diet. Kanebo puts the mail addresses in popular fashion magazines.

Cameras

The next step is to read the mail address or URL or even a bar code that designates the mail address or URL using a camera phone. This will further simplify user access and thus expand the use of magazines and posters to attract mobile members. Such new technologies will probably increase the value of services such as Sugu mail, depending on the firm's incorporation of the camera function in their service.

Several phones released in 2002 are "somewhat" capable of reading URLs with a camera. Users merely select the appropriate functions from the menu and place the URL between two lines that appear on the camera lens. URLs printed with raised type are easier to read than those without it. Better camera resolution and better pattern recognition will solve these problems by the end of 2004; the latter comes with higher processing speeds.

The biggest issue may involve users. Will people be willing to use cameras to access sites? The Japanese mobile and the U.S. PC Internet suggests that some people are very willing to adopt new technologies. And if they are willing to adopt this one, the URLs printed on posters, magazines, or even menus and napkins will become even more valuable. Very high resolution camera phones may be able to read very small bar codes that take up very little space in the magazine.

LINKING DATABASES

Many retail outlets have already linked their POS and membership card databases. This enables retail outlets to do data mining and analyze the purchases by individual users. The retail outlet can analyze these purchases in terms of age, gender, home address, and other data that are collected when people become members. Such an analysis supports direct mail and other forms of advertising. The next step is to link the mobile mail database with the membership card database. Unfortunately, as of early 2003, very few Japanese retailers had integrated these databases and thus were unable to take full advantage of the low cost and fast response times of their mobile mail services.

Retailers cannot integrate these mobile mail, membership, and POS systems unless they know each customer's mail address. And even if users register their mail addresses when they become members, many people often change these addresses for various reasons. In Japan, users must do this when they change service providers. This problem is exacerbated by problems with Spam.

In Japan, at one point, more than 90% of the mail sent to NTT DoCoMo mail addresses was Spam, thus causing mail boxes to be filled with Spam; worse yet, the users had to pay to receive this Spam. Most subscribers changed their mail addresses and some multiple times to avoid the Spam. The other service providers have had similar problems and since these problems are also widespread in Western PC mail services, it is likely that they will also occur in Western mobile mail services.

Jeansmate is the leader in linking the mobile mail and membership databases. During a purchase, the cashier checks the membership card and when discount coupons are presented, the cashier scans the bar code, which represents the discount coupon, as it is displayed on the phone screen. Either by reading the bar code (which contains a unique user ID) or by inputting the member's number, information about the member is displayed on the register. This includes information on whether the member has registered their mail address or if mail sent by Jeansmate has been returned in recent mail transmissions. In either case, the clerk will ask the customer if they would like to register their mail address with Jeansmate using a new technology called Xnavi. This device activates the mail function in a mobile phone,[2] causing mail to be sent to Jeansmate. The cashier merely inserts the Xnavi device into the port that is used to upload software as one of the final steps in producing a mobile phone.

These new technologies are helping Jeansmate update mail addresses and sign up mobile members in its store. As for the former, Jeansmate expects to update the approximately 70% of mail addresses that were unusable. The latter reduces the need to sign up members using other means that were described above, such as paying for the advertising services offered by the service providers. Jeansmate hopes to gradually obtain the mail addresses for its major customers within its 1.7 million members.

The updated mail addresses enable Jeansmate to continue strengthening its mobile mail services. The combination of its membership and POS system enabled it to know the age, gender, address, and spending level of each customer. Jeansmate had used this information to analyze purchases and target specific geographic areas with newspaper inserts. By knowing the mail addresses, it can also target users with mobile mail that is based on purchasing history and develop one-to-one relationships with them.

Jeansmate is also making organizational changes in order to more effectively use these mobile mail services. While previously its corporate information group had devised one type of mail message for all users, it now plans to work with the stores to develop mail messages for different stores and users. Together they will define the different categories of users.

New technologies will continue to change these services. Java is expected to have the largest effect on these mail services. Mail can activate a Java program that was previously sent to a phone. One form of Java program that Jeansmate would like to send is a screen saver that contains the Jeansmate logo. Thus the screen saver can function both as an interface between Jeansmate and the user and as a form of advertising for Jeansmate. Like other firms, Jeansmate has been sending free ringing tones to attract mobile members, but it may start sending free Java-based screen savers in the future.

PHONES AS MILEAGE AND MEMBERSHIP CARDS

Phones can also be used as mileage/point cards, which will likely be convenient for users and probably expand the number of people who will use these types of cards. In Japan it is not uncommon for some people, particularly young women, to carry more than 50 cards in their handbags. Supermarkets, restaurants, retail outlets, and stores have created mileage/point-card systems. Some of these cards are magnetic cards that are run through a magnetic reader, while others are paper cards in which the imprint from a special stamp indicates the number of points.

One method of using phones as point cards is to read a bar code that is displayed on the phone screen; Jeansmate is doing this to combine the mobile and membership databases. Furthermore, more than 10 Japanese firms announced ASP-type services in the first four months of 2003. In this case the only problem is the user's inability to confirm their number of points, a problem that also exists with most magnetic cards. Ideally, users would be able to access this information free of charge at any time. One solution is for the cashier to provide this information when a member makes a purchase, perhaps in the form of a printed value on the receipt. Another option is to allow users to access this information on the Jeansmate site. But this requires users to incur packet charges.

Infrared Connection

A more sophisticated solution uses an infrared connection to exchange data, including the authorizing of user membership and updating the number of points. Geo, Japan's second largest video rental chain, has introduced the appropriate readers, which cost less than $100, in its almost 500 video rental stores. Members register on Geo's PC or mobile site (the URL is advertised in each Geo store) by first downloading the appropriate Java program and inputting their membership number and password. In the stores, they merely press the appropriate button on the phone to activate the Java program as they hold the phone near the scanner; this authorizes the user's membership and updates the point card data.

The two advantages of infrared technology over bar codes are directionality and two-way communication. The line-of-sight requirement increases security while two-way communication enables stores to update the number of points in the phone (and the user can access this at any time) and to send other product information during and after such a session. For example, Geo is planning to use the Java program to remind users about overdue videos, manage online reservations, and send discount coupons.

A potential disadvantage of infrared is its speed. In Geo's implementation, it takes about five seconds for user authorization to be carried out, which is too slow for most retail outlets. On the other hand, infrared specialists such as Link Evolution have demonstrated connections that take less than one second for authorizing credit card numbers.

Both the infrared and bar code methods will create a need for supporting services. While some individual supermarkets, restaurants, retail outlets, and stores can do this with their own resources; there are many smaller places of business that do not have such resources. This is particularly true for those stores that either have not implemented membership systems or if they have, they distribute paper cards in which the imprint from a special stamp indicates the number of points. This suggests that there is an opportunity for firms to provide an ASP service for small restaurants and other businesses.

New Applications and Users

There are a number of other applications for these infrared solutions that will enable phones to be used as business cards, credit cards, and as remote control units in conjunction with other devices. If the information printed on business cards were stored in phones, the infrared function would allow the easy exchange of this information, thus allowing us to always access the information from our phones. An intermediate step is to include the information in bar codes that are printed on the business cards and to read them with a camera phone. Such phones already exist and one of these phones (from J-Phone) includes a function for editing this information in the phone's address book.

Phones can also be used as remote control devices with printers, televisions, karaoke devices, and vending machines. Convenience stores such as Lawson have made it possible for customers to print out information on their mobile screens using the infrared function and in-store terminals. It is possible to save the karaoke song codes in phones and subsequently request them on your next visit. NTT DoCoMo's c-mode service enables items to be purchased from a vending machine with a mobile phone, and one method of doing this uses the infrared function.

It may also be possible to use the infrared function for phones to interact with various devices on a person's body. This includes wearable computing, in which various types of clothing are used as displays (e.g., watches) or battery chargers (e.g., shoes). Another option is for phones to monitor medical devices that are implanted on the body and send the data to doctors or ambulances when necessary. It is also possible to use phones as tickets, and even money. As we shall see in Chapter 9, there are a number of other methods including 2D bar codes and smart cards. Bar codes may be cheaper but they are slower and probably less secure than smart cards.

These new technologies will strengthen the role of the phone as the new contact point for customers. As business people begin to use phones as business cards and to download information to printers in convenience stores and airports, it is likely that business users will begin to use phones for some of the other applications that are described in this chapter. Just like user learning played an important role in the early growth of entertainment applications and mail in the Japanese mobile Internet, user learning will also play an important role with business users. This has important implications for suppliers of customer relationship management and software, many of whom have been ignoring the mobile Internet both in Japan and elsewhere. Like many disruptive technologies, the mobile Internet has started with an unexpected group of new users, young people, but it may end up diffusing to business users.

SUMMARY

The mobile phone is becoming the new contact point for customers. Retail outlets and manufacturers are using the mobile Internet and SMS to send discount coupons, conduct surveys, offer free samples, and improve their brand image with young people. New technologies such as 2D bar codes, short-range infrared, Java, and devices that activate a phone's mail function offer additional ways for retailers to develop stronger relationships with customers, including the use of phones as mileage/point cards. While the initial users of these applications and technologies are young people, some of these technologies such as infrared technologies will probably expand the set of users to include business users.

NOTES

[1] See Jamie Cattell, Vice President, Research International, "M Research in Japan: The Mobile Internet Revolution and its Implications for Research."
[2] As of late 2002 it only worked with phones from NTT DoCoMo and KDDI phones and not with those from J-Phone.

Chapter 6

Mobile Shopping and Multichannel Integration

Chapter 6 continues the argument started in Chapter 5 - that the mobile Internet is starting a new reorganization of the customer's purchasing process. Like the PC Internet, the mobile Internet will not replace the so-called "old" economy in the foreseeable future. In fact, it has less of a chance to do so than the PC Internet, given the small size of their screens and keypads. But just as the mobile Internet is changing the retail industry, it is also changing Internet shopping in unexpected ways. The products, customers, and purchasing methods are different on the mobile Internet than on the PC Internet, and a set of different technologies will drive the expansion of the market. Mobile shopping in combination with magazines is already the fastest growing part of the Japanese mobile shopping market and in the long run the integration of the mobile phone with other media such as magazines, the radio, and the television will probably drive the expansion of mobile shopping.

Figures provided by service and content providers suggest that about $267 million in physical products were purchased on the mobile Internet in 2002 or 2.7 times the amount purchased in 2001. The number of accesses from the official i-mode menu and data from content providers give us a rough snapshot of those products that are sold over the Japanese mobile Internet. Table 6.1 summarizes the order of traffic by subcategory, and it lists firms in each subcategory. The leading category is CDs, DVDs, games, and books followed by convenience stores, concert tickets, and fashion. Based on interviews with several of the firms in each category, it appears that the CD and DVD sales dominate the sales in the first two subcategories and thus mobile sales in general followed by concert tickets (dealt with in Chapter 8) and fashion. The popularity of CDs, DVDs, concert tickets, and fashion is quite different from the leading sellers in the U.S. and Japanese PC Internet, which are travel, computers, books, and software.

The greater importance of CDs, DVDs, concert tickets, and fashion-related products in the mobile Internet than in the PC Internet reflects the small screens and young mobile Internet users and thus the disruptive nature of the mobile Internet. It is difficult to search for, choose, and pay for more complex products like personal computers and airline tickets on the mobile phone. It is also difficult to search for CDs, DVDs, and fashion-related products on the mobile Internet; this is why most of these products are currently selected from mail services as opposed to search engines. Furthermore, young people are the major users of the Japanese mobile Internet and European SMS markets and the products that are being purchased on the Japanese mobile Internet reflect the strength of the youth market. Shopping sites that sell products on both the PC and mobile Internet also report much lower ages for the buyers on their mobile sties than on their PC sites.

These differences in products, customers, and purchasing methods are major reasons why Japan's leading mobile shopping sites are quite different from its leading PC shopping sites. Not only are CD and fashion sites doing better than travel and book sites, there are also different winners within the same category. For example, within the category of CDs and DVDs, HMV has done better than Tsutaya Online in the PC Internet, and the opposite is true in the mobile Internet. Similarly, within the category of virtual malls, Rakuten has done better than Net Price, and the opposite is true in the mobile Internet. In both cases, the successful mobile shopping sites introduced the appropriate mail and other services while the less successful sites have been slower to introduce these innovations. Similarly, new entrants like Xavel and Index have become shopping leaders in the mobile Internet by introducing innovative services that match the users and characteristics of the mobile Internet.

Also unlike the PC Internet, many of the leading mobile shopping sites have been profitable for several years. While most PC Internet shopping sites have attempted to compete with retail outlets in terms of price, mobile sites primarily compete in terms of convenience and brand image, thus providing margins that are similar to those found in the retail industry. Fashion-related products have margins that can be as high as 30% while products like CDs and DVDs have lower margins. This suggests break-even points of between $85,000 and $255,000 a month[1].

The biggest questions for mobile shopping are how it will evolve outside of Japan and how improved technology can change and expand this market. I believe that those young Americans who spend little time in front of their PCs will react in much the same way that their Japanese counterparts do. As for technology, faster processing speeds will increase the size of Java programs and enable other more advanced user interfaces. The use of camera phones can facilitate the integration of the mobile phones with other media such as magazines, and new forms of infrared or W-LAN may do the same thing for televisions. This chapter first discusses these issues for pure online shopping followed by the integration of the mobile phone with radio programs, magazines, and TV programs.

PURE ONLINE SHOPPING

Most of Japan's leaders in mobile shopping currently pursue a pure online strategy. Most of the sites shown in Table 6.1, particularly those in the top category of CDs, DVD, Games, and Books, currently do this. Although many of them sell products on both mobile and PC sites, they sell a different type of product and to a different user and in a different way with their mobile sites.

TABLE 6.1. Order of Traffic by Sub-category and Firm in the i-Mode Official Shopping Category

Subcategory	Leading Content Sites and Firms
CDs, DVDs, games, and books	Tsutaya Online, Music Shop/Vibe, HMV Japan, Kinokuniya, Honya-san, Book Service, Amazon.com, Playstation,
Convenience stores	Lawson, Family Mart, 7-11
Concert tickets	Ticket Pia, Lawson Ticket, E-Plus
Fashion	Accessory (Index), Perfume (Index), Uniqlo, Brand Love, Sutailief, f-mode, smart & mini, Image, Shibuya 109, Magaseek
Flowers	i-Flower Shop, Flowers (Index)
Lifestyle	Net Price, Senshukai, i Dinos, Wine

Music and Videos

We learned about Tsutaya Online's success in using the mobile Internet as a new contact point for its retail operations in Chapter 5. Partly through this success, it has become one of the leading pure online mobile shopping sites on the i-mode menu. It had sales of about 3.0 billion yen between April 2002 and March 2003, or more than twice its sales on the PC Internet in spite of similar start dates. DVDs and CDs each represented 30% of the sales, and the rest was in games, books, VHS videos, and various movie paraphernalia. And while the average age of a buyer was 30 on the PC site, it was 21 on the mobile site.

More interestingly, few people buy these products by using Tsutaya's mobile search engine. Instead, 60% of the sales are through purchases made after viewing mail and the second and third leading methods are buying products shown at the top of the page (recommended products) and in product rankings. Many of these items are purchased before they are released using an advanced order service.

Tsutaya Online was sending about one million mail messages every day to its 2.4 million members in mid-2002. Some of the mail is sent to all members while most is sent to people who registered for one of the 50 types of mail offered by Tsutaya Online. Many of these different mail types reflect different types of music, movies, and books. They contain entertainment, product information, store information, and/or advertisements. While PC click rates and fees have dropped dramatically in the last year, this has not happened in the mobile Internet, and Tsutaya Online's income from advertisements in its mail messages is similar to its online sales

HMV, also a large music retail outlet in Japan, has found similar results in the pattern of its mobile sales. The mobile Internet does represent a smaller percentage of its sales than for Tsutaya Online; only 22% of them are from the mobile phone. This is partly because HMV has not used mail services to the extent that Tsutaya has in both shopping and in retail. The lack of the former mail services makes it harder for visitors to its site to make purchases, and the lack of the former has made it harder for it to create a mobile portal that is as strong as Tsutaya Online's portal.

However, as shown in Table 6.2, mobile users like to buy Top 40 music on the mobile Internet just as Tsutaya Online's customers do. While the advantage of the PC Internet is search engines, which enable purchases of so-called deep catalogue music, the advantage of the mobile Internet is portability and easy purchases of products while hanging out with your friends. It seems like many young people just want to

TABLE 6.2. Type of Music Sold by Channel for HMV (2002)

Type of Music	Stores	PC Internet	Mobile Internet
Pre-release	0	30%	35%
Top 40 and back catalogue	98%	40%	60%
Deep catalogue	2%	30%	5%

obtain the most popular music when or if they can before their friends do.

Although HMV does not collect data on how many products are purchased in mail, new releases, or rankings, it did provide an interesting logic for the popularity of them and top 40 music. According to a representative of HMV, young people tend to place a much greater importance on music than do older people, and they also see it as a social as opposed to a private phenomenon. They want to purchase popular music so they can fit in with the main social groups. This encourages them to choose music from rankings and new releases. On the other hand, as people age, they typically reduce their emphasis on music and see it as a private phenomenon. Most record companies find it much easier to target these young people and still don't know how to reach the mature users.

Another music-related service that will likely succeed is a concert goods shopping service. Many concertgoers are unable to purchase T-shirts and other band-related paraphernalia after a concert due to the large crowds at these concerts. Firms like Cybird have started sites that offer these products in cooperation with the firms that sell these goods at the concerts. A sales booth at the concert lists the URL; and in the joint ventures that involve Cybird, Sugu mail is used. Users merely send an empty mail to a designated address and quickly receive a response containing the appropriate URL. These sites have experienced strong sales in the several hours following a concert.

Fashion

It is quite likely that fashion-related products represent a larger percentage of on-line mobile sales than are suggested by Table 6.1 for several reasons. First, the two leading sites listed under lifestyle mostly sell products that can be defined as fashion. More importantly, several of the leading sellers of fashion-related products are, or at least were, unofficial sites until recently. For example, Girls Walker is one of the leading suppliers of online mobile shopping services, but it is still an unofficial site. Furthermore, many of the currently successful official sites on the shopping menu (e.g., Net Price and Senshukai) first succeeded as an unofficial site and thus still receive most of their accesses through bookmarks as opposed to browsing through the i-mode official menu.

Xavel has used its strength in mail magazines to also create a successful shopping site within its Girl's Walker portal. It started selling products on its site in late 2000, it had 1.2 billion yen (US$ 10 million) in shopping sales in 2002, and by April 2003 it its sales had reached 200 million yen (US$ 1.67 million) per month. While initially it had to assign a more than half its employees to finding products, its success has caused the manufacturers to come to Xavel. For example, it now works with a fashion show called Kobe Collection and several fashion magazines.

Most of Xavel's sales come through its mail magazines. Advertisements for various fashion-related products are included in the mail, which include a URL. This is facilitated by the fact that many of the mail magazines provide information about some aspect of fashion. This increases the chances that an advertisement will lead to a sale. Of course, there is a huge opportunity for Xavel in the area of numerical analyses. Which products are most appropriate for each magazine, which magazines have the highest sales, and how can Xavel move women from magazines with low sales to those with high sales are some of the analyses that Xavel can and should carry out.

Index has also become one of the leading providers of mobile shopping services, although it has primarily done this through alliances with other firms. Two of its sites provide pure online shopping services while several others, which are described later, are done in combination with magazines and TV programs. The two sites that provide pure online shopping sell perfume and flowers. The perfume site has done the best with almost 200 orders per day at 5000 yen an order.

Both sites depend to some extent on mail services. As of late January 2002, Index had 300,000 registrants for its perfume and flower sites (official i-mode sites), with the majority being with the perfume site. The unique part of the flower site is that it offers recommendations for flowers based on the event (e.g., going-away party) or birth date, and users can register to receive mail right before a friend's birthday as a reminder to send flowers. The unique part of the perfume site is information about the perfume used by popular actresses. Shoppers can search for their favorite actress and purchase the perfume used by the actress. Information about the habits of popular actresses is found in the public domain (e.g., in magazines).

Net Price may be the most successful of all mobile shopping sites. Net Price has long been the second leading provider of virtual shopping malls on the PC Internet. Like the leader, Rakuten, Net Price charges a fixed fee and a percentage of their sales to be on Net Price's virtual mall and use its technology. Unlike the leader Rakuten, however, Net Price quickly created a new business model for the mobile Internet that makes use of the popularity of mobile mail with young people and in some ways is creating virtual communities of mobile shoppers.

Net Price periodically sends mobile mail to its members, offering them unsold brand name products such as clothing, ladies handbags, watches, jewelry, and other accessories at a price that depends on the quantity ordered. Members are given one week to attract multiple buyers and thus obtain a larger discount; the maximum discount is typically about 30%. Many members use mobile mail to gather a large number of buyers. As of late 2002, Net Price had 2.6 million members and had sales of more than 300 million yen in March 2003, up from 80 million in June 2001.

Manufacturers or retail outlets see Net Price's service as another way to sell unsold inventory. For many of these manufacturers and retailers, it is better to sell them cheaply on the mobile Internet than in stores where the discounted prices can degrade the value of all the products. As discussed in the next section, mobile sites such as Magaseek are also considering the use of the mobile Internet in combination with magazines as another method to sell unsold stocks.

Book Sites

Books are not selling any near as well as CDs and videos on the mobile Internet. For example, the leader is Kinokuniya and it was only selling 15 million yen in books per month in the summer of 2001, which was only 5% of its total online sales. Even if the number of sales may have doubled since then, this is still far below the sales of Tsutaya Online and Xavel.

In fact, Tsutaya Online may sell almost as many books on the mobile Internet as Kinokuniya, but the types of books are completely different. Tsutaya Online sells books about entertainers who are popular with young people; young people purchase them just like they purchase CDs and videos on their mobile phones, after receiving mail or when looking through book rankings or new releases. The strength of Tsutaya Online in books reflects the disruptive nature of the mobile Internet and its effect on the main users and the method of shopping; most people expected Kinokuniya or Amazon to dominate sales of books on the mobile Internet.

Amazon.com has changed the Japanese PC Internet book market as it has quickly overtaken Kinokuniya, which is Japan's leading book retailer. Unfortunately, it was a late entrant to the mobile Internet, and as of early 2003 it still lagged several book sites such as Kinokuniya, Honya-san, and Book Service in mobile sales. The bigger problem for Amazon and other mobile book sites is the difficulty of searching for books on the small screens of mobile phones.

Convenience Stores and GPS

Many people believed that convenience stores could become the places where consumers, who are never at home or can't receive deliveries at home, pick up the products they ordered on the Internet. While it may still happen, it hasn't quite worked that way yet in Japan. For example, one convenience store's mobile sales only reached 43 million yen in February 2002, down from a peak of 72 million yen in December. Its PC Internet sales were somewhat higher than this, and it is possible that both the PC and mobile Internet will continue to see growth. But if growth does not occur in Japan where there are more convenience stores per capita than in probably any country in the world, it is unlikely that it will catch on elsewhere.

One alternative to convenience stores is GPS. Historically the post office and other services have delivered mail to addresses and not people. GPS provides post offices and firms with the possibility of sending mail and other packages to a person's actual location. This is clearly not a service for everyone. But some people either want to receive something very quickly, they are rarely at home, or they cannot receive deliveries at their office. People will probably have to pay a premium and perhaps provide some location information in addition to their GPS location in order to receive such deliveries.

Java and Other Technologies

A better user interface is needed to expand pure mobile sales. The small keyboard and display make it hard to search for products, and this makes it difficult for content providers to offer a full range of products. Selling top 40 music and perfume that is used by popular actresses is a limited market.

Some people believe that Java will provide the first such improvement in user interface. For example, K-Labs, the leader in Java applications in Japan, offers a so-called "thumbnail" viewer. This Java program displays pictures of about six products on one screen. Users can see select products based on pictures rather than text. And because most of the information is contained in the Java program, only small amounts of data must be downloaded. Rakuten has used this function and an increase in products and stores to dramatically increase its mobile Internet sales. While it is still behind Net Price, its sales reached 220 million yen in March 2003.

Other technologies may lead to a further expansion of the mobile shopping market. By combining a Java program with larger displays (e.g., based on light-emitting polymers) and/or software that renders 3D images in real time, it may improve the user interface enough to enable pure online shopping to become a significant market. Three-dimensional images enable users to rotate products on the screen; this is a function that might be useful when looking at fashion-related products. It might also be possible to use the camera function to input your body dimensions and thus compare them with the 3D image of the clothing.

CATALOGUES AND MAGAZINES

A more promising approach in the short run is likely to be the combination of mobile phones with other media such as catalogues and magazines. Catalogues and magazines provide much larger pictures than can be displayed on a mobile phone screen and more importantly they allow mobile sites to tap into established consumer behavior. Already more than 50 magazines offer mobile shopping services largely in combination with other firms.

Catalogues

Most people look at catalogues with at least some intent of ordering something. And apparently a lot of people do, because catalogues have evolved into a multi-billion dollar industry since Sears Roebuck first introduced them in the nineteenth century. Consumers have become accustomed to looking through catalogues; Sears Roebuck, L.L. Bean, Lands' End, and others have learned to design these catalogues in a way that facilitates searches and purchases.

Some people expect the PC Internet to replace these catalogues as consumers become accustomed to using the Internet. Filling in forms and placing orders over the phone may be a nuisance to some people, particularly those accustomed to the PC Internet where response times are generally fast and registering credit card information facilitates orders. On the other hand, many consumers do not like the requirements that the PC Internet imposes on them. They must get up from the couch, which is one place where they like to read catalogues while they watch television, and sit down in front of the PC.

I believe that the mobile Internet complements consumer behavior more than the PC Internet. With the mobile Internet, consumers can still relax on the couch, on the bed, or even in the bathtub while they look at the catalogue and place their orders. And since many people have their phones with them at all times (yes, even in the bathroom), the required changes in consumer behavior are much smaller for the mobile Internet than for the PC Internet.

The leading catalogue in Japan is operated by a firm called Senshukai. It entered the mobile Internet as an unofficial site in January 2001, and from late 2001 it has operated an unofficial site (Belle Maison) on the i-mode menu. In spite of starting its mobile Internet site several years after starting its PC Internet site, mobile Internet sales represented almost 25% of total Internet sales by the spring of 2002. More importantly, the mobile Internet has attracted younger users. While the average age of the catalogue users is 31, the average for the mobile users is 27.

Triumph, one of the world's leading providers of women's underwear, is also trying to combine its catalogue with the mobile Internet. Unfortunately, Triumph is trying to establish both a new catalogue and a new site (it had 120 million yen in sales in 2002). Senshukai is merely trying to add a mobile dimension to its already successful catalogue while several other firms are trying to do this with successful magazines. Adding a mobile site to an already successful service is much simpler than creating both a site and a successful catalogue or magazine.

In any case, combining the mobile Internet with established catalogues probably has much more potential in the United States than in Japan. Catalogues are a recent phenomenon in Japan and in fact were first started by U.S. firms like L.L. Bean and Lands' End. In the United States, buying clothing and other products from catalogues has become a way of life for many Americans. And many of these Americans will likely find the ability to make these purchases from their mobile phone more convenient than doing it from a PC or making a phone call.

Magazines

Mobile Internet sales in combination with magazines probably represent a much larger market than for those in combination with catalogues. Most consumers, even in the United States, read far more fashion magazines than catalogues, and thus sites that are integrated with magazines can tap into a much richer consumer behavior than those sites that are integrated with catalogues. Consumers are accustomed to searching for magazines in bookstores and searching for information in those magazines. Bookstores and magazines are designed to facilitate these searches and in particular facilitate the purchase of magazines and the viewing of and acting on advertisements.

In the long run we may see a new business model emerge for magazines as some brand name advertisers begin to expect more than just a pretty picture of their clothing, cosmetics, or accessories in a magazine. They may begin to expect Internet sales from these advertisements, and a portion of these sales may become a significant part of a magazine's income.

One firm that believes this is Index. Index acquired a publisher of youth-oriented magazines in 2002. One of these magazines features silver jewelry for men, admittedly a special market, but exactly the kind of market that is difficult to reach with a pure online mobile shopping site. Index began selling the jewelry that is advertised in these magazines on a mobile site in September 2002. The site had 2500 orders in the first six weeks and had jumped to the top of the fashion category in terms of accesses on the i-mode menu.

However, the leader in combining a mobile shopping site with magazines is Magaseek, which is owned by one of Japan's largest trading companies, Itochu. Magaseek began selling products that are advertised in magazines in August 2000 as an official i-mode site. It focused first on magazines since surveys have found that 80% of young people find fashion information in magazines as opposed to 50% on TV, which was the second largest source of information. It started with those magazines that offer the most fashionable clothing and accessories and organized the site by magazine type, issue, and page.

Magaseek had 100 million yen in sales in November 2002 up from 30 million yen in August 2001. Fifty-five percent of the purchases were from the mobile Internet as opposed to the PC Internet, which was down slightly from 70% in August 2001. Of these sales, Magaseek takes the same margin (30 - 40%) as do upscale boutiques and department stores in Japan, and the prices are the same as in these upscale stores.

One advantage of Magaseek's (or Index's) business model is that it does not have to advertise its products or carry inventory. Magazines provide the advertising and database for the service since the order process is described in each issue. Magaseek outsources the distribution function to another company, which combines products into a single order and then contracts with a delivery company.

The challenge for Magaseek is to increase the number of sales without substantially increasing the search time for users. Magaseek needs to increase its sales volumes in order to obtain volume discounts from manufacturers. One option is to take unsold products from upscale boutiques and department stores and sell them at 70 - 80% of the original price. Currently, these unsold products are sold at 30 - 40% of their initial price in discount shops, and thus a site like Magaseek could provide higher revenues for the clothing manufacturers than discount shops.

Magaseek needs to do this without increasing the search time for users. In the current hierarchical search process offered by Magaseek, increasing the number of items offered on the site would increase the search time. Printing the URL next to the product might speed up the search process since users would no longer have to search through Magaseek's menu. On the other hand, inputting the URL can be time-consuming, and Cybird's Sugu Mail is one option.

Another option is to use the camera phone to automatically access the URL as is described in Chapter 5. If users become accustomed to this approach, it is possible that the value of advertisements in fashion magazines will change substantially as a more direct connection is made between the advertisement and sales. Advertisers may get a better idea of which advertisements in which magazines are the most closely connected with sales.

A more radical approach is for users to access mobile shopping sites through their phones and RFID tags. As these devices become commonplace in clothing and other products, phone manufacturers may place the appropriate readers in phones. This would enable consumers to access (perhaps even secretly) information about another person's clothing that could include the URL for a mobile shopping site. After all, many people probably make portions of their purchasing decisions as they look at the surrounding clothing and other products. Furthermore, users may also want to access information (such as ingredients) about products in stores or other places before they make a purchasing decision.

RADIO STATIONS

The mobile Internet is becoming an important part of the total entertainment package of radio stations. A participative environment has always been an important part of radio stations, and both the PC Internet and mobile Internet increase the level of consumer participation possible in radio station programs. Many Japanese radio stations have been receiving more music and concert ticket requests from mobile mail than from any other source since mid-2002.

For example, the leader mobile contents providers among radio stations is FM 802 in Osaka. While there were very few requests from mobile phones or even PCs in the year 2000, mail from mobile phones and PCs each represented one-third of total concert requests by mid-2001 and about half of music requests. By early 2002, mail from mobile phones was the leading source of both concert ticket and music requests for FM 802 and other radio stations such as FM Yokohama.

FM 802 and others encourage listeners to access their PC and mobile home pages and submit requests via the Internet through their point systems for members. FM 802 had 13,000 members in its Internet program in September 2001, and it gives points to members each time they make a request, access their home page, or respond to a survey on the Internet. Members can use these points to obtain free concert tickets.

FM 802 encourages its listeners to use the Internet because FM 802 wants to provide its sponsors such as concert and event promoters with additional exposure to its young listeners on its PC and mobile home pages. For example, in late 2001, FM 802 promoted the completed renovation of an old warehouse district into an entertainment area that includes a variety of music clubs. Users could listen to sample music by bands scheduled to play at these clubs by clicking on the appropriate place on the home page.

Another option for radio stations is to link their sites with CD shopping sites. Many radio stations already offer information on their sites about the music being played on their radio programs and links between their sites and CD shopping sites. Although the sales are still low, radio stations have high hopes for such sales as the service providers approve of proper linkages between the various sites and other problems are solved. Another alternative is to link their sites with sites that provide music-downloading services. KDDI's successful vocal ringing tone service suggests that many people will be interested in downloading music onto their mobile phones if the price is right, and many people first hear a new song on the radio.

The biggest challenge for linking the act of listening to music on the radio with the purchase of the music is accessing a radio station's site. Magazines hope that camera phones will facilitate site access while consumers are reading magazines. If someone can come up with an equivalent way to use some of the information in the radio signal to facilitate access to a radio station's mobile site, it may be possible to expand mobile shopping in combination with radio programs.

TELEVISION PROGRAMMING

The integration of television with the mobile Internet probably represents a much larger potential market than the integration of radio with the mobile Internet. Television viewing and the television advertising market far exceed their counterparts in the radio industry. For many people television represents their main form of home entertainment and source of information.

The mobile Internet offers the possibility of interactivity to television, something that digital television was supposed to offer. Digital television has failed to materialize largely because the benefits of digital television are still unclear. The integration of the mobile Internet with television programming can provide us with some interactivity and more importantly help guide the television industry towards digital television.

Combining Japan's experience with the mobile Internet and Europe's experience with SMS may provide a better business model than either of them by themselves. Japanese television companies have became relatively successful providers of entertainment contents by using the theme songs, animated characters, and popular actresses and actors from their popular programs. On the other hand, European television companies have used the revenue sharing systems for SMS messages, which do not exist for mail services in Japan, to offer voting, contest entries, and quizzes with SMS. Future mobile Internet services should probably include a combination of entertainment and mail services along with the integration of these mail services and mobile Internet-based information.

Japanese Contents

Japanese television stations have already become leading providers of ringing tones and to a lesser extent screen savers, games, entertainment news, and mobile shopping services. For example, Fuji TV and Asahi TV are now two of the leading providers of ringing tones due to the popularity of the theme songs from some of their programs. Fuji TV, in cooperation with Index, offers several hundred songs from current and old TV programs. In early 2002 it had 200,000 subscribers, each paying 80 yen per month to download ringing tones. Both Fuji TV and Asahi TV have been successful with selling screen savers that are based on programs with animated characters or an actress or actor that is popular with young people.

Games are a second possibility. Fuji TV and Index began offering a game called "Joshi Ana" (female announcer in English) in December 2001, and the game obtained more than 20,000 paying subscribers (100 yen per month) within one month of its start. The purpose of the game is to become a famous announcer, and it complements the subject of the television program, which is the life of television announcers.

Early demand for mobile shopping services can also be seen. For example, Fuji TV and Index offer character and greeting card downloading services that are introduced in a children's program (Gotchappin) where the children's parents apparently buy the products. They have been offering Omiyageland since March 2001. Consumers can purchase products such as toy buses and cars on the mobile and PC Internet that are featured in a TV program called i-nori. The program is about young people who travel overseas seeking love and adventure.

European SMS

A key difference between Japan's mobile Internet and Europe's SMS services is that the European service providers share revenues from the actual sending of mail or messages while Japanese firms do not. Thus,

European television stations and other organizations that are able to induce consumers to send short messages can receive a portion (between 40% and 60%) of the SMS revenues (0.50 to 3.00 Euros per message) in Europe and elsewhere. Some of the main beneficiaries of these revenue sharing arrangements have been European television stations.

Some of these television stations have convinced consumers to predict the outcome of sporting events, vote in music contests, vote on alternative outcomes to TV programs, apply for news alerts, and participate in chat groups. It is not unusual for 200,0000 viewers to predict the outcome of a sporting event or choose an ending to a soap opera or a winning musical group. One game had viewer's answer how they would have responded to various conflicts that occurred in the show. Some users upload their name, hobbies, and photograph as part of on-screen chat where the other participants rate the photographs. A Finnish television station even allows participants to contact each other if both parties agree. These are all examples of firms creating and profiting from virtual communities. The popularity of these services has caused some European television stations to wonder if they will gradually lose control over their contents as interactivity increases[2].

Many Japanese television stations have experimented with these kinds of services, but the lack of a revenue sharing agreement on mail revenues gives them fewer incentives than their European counterparts. For example, Asahi TV has experimented with surveys and voting on several of their programs, typically those after midnight in order to reduce needed investments in servers. Currently they are looking for a way to increase ratings through these surveys and voting, clearly a much more difficult task than merely inducing viewers to send mail. Being able to receive a portion of the service providers' mail revenues would provide Asahi TV with a larger incentive to pursue these services. An alternative is for Asahi TV to provide these services only to paid monthly subscribers. Here getting people to pay before participating in the service might be difficult.

Future contents

Combining Europe's method of revenue sharing with Japan's mobile Internet services could further expand these services in both Japan and Europe. After all, clearly much more can be done with the mobile Internet than with SMS. For example, much more information can be input on a home page than in an SMS message. Users can select multiple candidates and choose from a list the reasons as they choose a winner or a candidate. News alerts would clearly benefit from a mobile Internet service where mail can include a URL and thus access to additional information. Services that ask viewers to choose alternative outcomes might include a game that dealt with these alternative outcomes.

Unexplored areas include information about the actors, actresses, restaurants, and other places that appear in the programs. Although Asahi TV currently offers some information about restaurants, nail salons, and hair salons that appear in TV programs, the traffic is very low. Finding the proper mix of services, business models, and information will take time.

Similar plus additional problems exist with providing viewers with the opportunity to buy products that are used in programs such as clothing and makeup. The sales of perfume that are used by popular actresses on Index's and Xavel's sites suggests a potential market exists. But finding the proper mix of services, business models, and information will take time, particularly when firms can easily waste a great deal of money and brand image on providing excessive amounts of information about clothing and cosmetics.

Furthermore, there are other potential obstacles that need to be overcome like the rights of actors and actresses. Actors and actresses expect to be reimbursed when clothing or cosmetics they wear/use in a TV program are sold. And currently, none of the TV stations appear interested in carrying out these negotiations, partly since these negotiations might possibly damage their existing relationships with actors and actresses. One needs to only look at the problems faced by firms trying to sell online electronic books to see some of the challenges faced by TV stations trying to provide shopping services. In an industry where image is critical, most TV stations will probably move slowly in this area.

New Technologies

A common problem for mobile Internet and SMS services is providing users with their mail or site addresses. European television programs display the SMS on the television screen, which some viewers apparently do not like. Although many Japanese television programs display facsimile numbers on the television screen when they are asking for comments, they have not yet started to display their mobile addresses.

New technologies that promote easier connection to mail or site addresses would be beneficial. It is already possible to use phones as a remote control device by simply placing the appropriate infrared technology in the phone. Information about the channels and the times in which they are accessed is saved in the phone and thus facilitates surveys of user viewing habits. It is also possible to choose programs on a mobile Internet site and use this choice to automatically change the television to the correct channel.

It may also be possible to have two-way communication between phones and televisions through the iRDA protocol that is discussed in Chapter 5. These devices have been in NTT DoCoMo's phones since mid-2002 and it is possible to put them in televisions (they cost less than $10). Although it is difficult to send these infrared signals from a phone to television (due to the low power of phones), some firms, such as Link Evolution, argue that it is possible to use the interference between the television and phone signals (similar to radar) to exchange information. Other options include Bluetooth and WLAN.

SUMMARY

The disruptive nature of the mobile Internet has caused the successful products, customers, and selection methods to be different from that of the mobile Internet and these differences have led to the success of a new set of firms. These new firms along with many incumbents are now attempting to integrate mobile Internet services with other media such as magazines, radio programs, and television programs. Larger and more complex Java programs, bar codes, cameras, and other technologies are expanding the possibilities of this multichannel integration.

APPENDIX: PAYMENT SERVICES

Most purchases are currently done with credit cards and cash-on-delivery (COD) (see Table 6.3) Although credit usage is lower in Japan than in the United States and many European countries, it is growing quickly and JCB, the leading credit card company in Japan, expected the percentage of online purchases (including the PC Internet) made with credit card companies to double to 60% in 2002. Of course, college and high-school students have a lower ownership of credit cards than those employed in full-time jobs in Japan and other countries. Young mobile shoppers probably use COD more than credit cards; surprisingly the mobile shopping sites report returns of less than 1% with COD.

Sites with retail outlets like Tsutaya Online, Lawson Convenience Stores, and HMV also offer store pickup services where payments are often made in combination with the pickups. People who live alone or in homes where no one is at home during the day are probably the main users of these services. Convenience stores either in their own mobile Internet services or in combination with other mobile and PC Internet sites are aiming their services at these kinds of people.

Banks are not doing very well in the mobile payment area. Most Japanese banks offer the same form of bank transfer services in the mobile Internet that they offer on ATMs. They charge between 105 and 630 yen to send a bank draft depending on the amount of money sent. Furthermore, Japan Net Bank, a joint venture supported by some of Japan's largest firms including Mitsui-Sumitomo Bank and NTT DoCoMo, offers services with fees as low as 10 yen for money transfers using i-mode phones.

Unfortunately, the banks' mobile services are nowhere near as popular as their ATM services. While money sent from an ATM is the most common method of paying bills in Japan, none of the mobile shopping sites reports payments from mobile phones and only one site, Pia, reports any payments from banks. Pia is somewhat unique in that it gives its customers a few days to pay (either by bank draft or by visiting a Pia

TABLE 6.3. Percent of Mobile Payment by Type of Payment

Firm	Product	Credit cards	COD	Bank	In-Store
Tsutaya	Music/Movies	40%	20%		20%
Pia	Tickets	50%		50%	
Lawson	Music/Movies				mostly
Magaseek	Fashion	50%	50%		
Cyber Wing	Music	>90%			
HMV	Music	58%	30%		12%
Index	Perfume/Flowers	15%-25%	75%		

Source: Interviews with the content providers in 2001 and 2002

ticket outlet) after they have ordered a ticket whereas other sites want payments to be made simultaneous with orders or when receiving the products.

I believe that the mobile payment services offered by banks will not succeed until the banks make these services easier to use. The banks currently require users to fill out a several page application that can only be submitted by regular mail. An even larger problem is that to send a single payment, users must input three ID numbers and three passwords, something that is very difficult and time-consuming to do on a mobile phone. Japan Net Bank's system is not much easier to use. Its service requires users to input two three- to eight-digit passwords and four digits from a 16-digit identification code.

Contrast this with credit cards where users merely need to input their credit card numbers just as they do on the PC Internet. And the mobile shopping and service providers have been making strong efforts to further simplify purchases with credit cards. Most mobile shopping sites enable their customers to register their credit card numbers and addresses to simplify future purchases. For example, J-Phone introduced a wallet service last December in which users only need to input their credit card and address information once, and after that they are offered access to this information when they try to make a payment on any official J-Phone site. According to J-Phone, more than 30,000 users had registered their credit card and address information by early 2003.

NOTES

[1] As discussed earlier, it is generally agreed that firms need 30,000 subscribers at 100 yen ($0.85) a month to break even. For margins of 10% to 30%, this requires between $US 85,000 and 255,000 per month in sales to break even.
[2] "Mobile industry looks to SMS/TV interactivity," *Europemedia.net*, Feburary 18, 2003.

Chapter 7

Navigation Services

Many people consider the navigation market to be the largest potential market for mobile Internet services. This is largely due to the portable nature of mobile phones, the potential use of position information from GPS (Global Positioning Systems) and base stations, and the success of car navigation systems (which use GPS). Car navigation systems are particularly popular in Japan where there are more than 10 million installed units (See Figure 7.1), or about 14% of the 73 million vehicles used in Japan. The market for such systems has also grown quite large in Europe and similar trends are also emerging in the United States where more than 300,000 were sold in 2002, up from 175,000 in 2001[1].

However, translating these navigation functions from a several thousand-dollar device to a sub 100-gram mobile phone or even a PDA that is expected to cost less than $400 is a very difficult task indeed. The car navigation systems have much larger and higher quality color displays than mobile phones, and they store the maps on hard disks, on DVDs, or, in the case of the less expensive systems, on CD ROMs. The smaller screens and memories make phone- or even PDA-based navigation systems a disruptive technology when compared to car navigation systems.

The disruptive nature of the mobile phone has caused a new set of applications and users to emerge in the Japanese mobile Internet. Instead of phone-based personal navigation systems that help people navigate Japan's crowded streets via detailed maps and GPS technology, simple train and restaurant information services dominate the naviga-

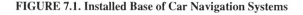

FIGURE 7.1. Installed Base of Car Navigation Systems

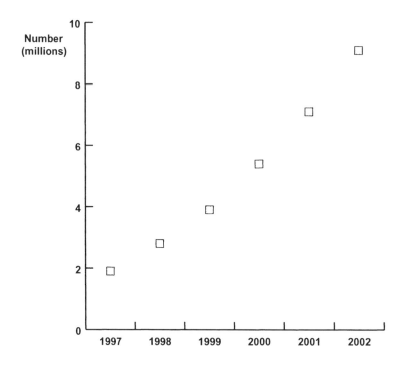

tion market. The mobile Internet is much more disruptive for the map providers than for the train and destination information providers because the small and poor resolution displays impact more on the map than on the other two services. This is why the train and destination service providers entered the market earlier than the map providers and they may have more than 50 times the traffic (in both cases) and paying subscribers (only in the case of the train services) than other navigation-related services in Japan's mobile Internet (see Table 7.1).

The concept of disruptive technologies also explains why GPS-enabled phones are attracting a new set of users that are different from those that use the conventional GPS systems. GPS-compatible mobile phones are lighter and cheaper[2] but incur much higher packet charges than conventional GPS systems in order to overcome the power consumption problems in phones. Thus, a new set of workers, people that commute by train and bus, are the initial users of these GPS-enabled phones. Of course, technological improvements will continue to reduce the packet charges of the GPS-enabled phones and may enable mobile phones to eventually replace the conventional GPS systems that are used in existing commercial applications like fleet management in the future.

This chapter analyzes the mobile Internet navigation market in terms of disruptive technologies and network effects where network effects play a strong role in destination information services. This chapter discusses these issues first for Internet-based car navigation systems where Internet connections via mobile phones have enabled a new form of car navigation system to emerge. This is followed by consumer services such as train, bus, map, and destination services, the integration of these consumer services, the integration of B2B and B2C services, business services, and finally location-based services.

Some of these services are more relevant for the United States and Europe than for others. The Internet-based car navigation systems are clearly a possibility in the United States in spite of the fact that car navigation systems are still not widely used in the United States. On the other hand, the lower population densities in U.S. urban areas make train and bus information systems and even restaurant services less important in the United States than in Japan or other Asian and even

TABLE 7.1. Ranking of Navigation Contents by Traffic and Order of Entry Traffic

Ranking	Type of Contents	Number of Contents Available at the Start of i-mode	Number of Paid Subscribers
1	Train routes	2	Almost one million
2	Restaurants	4	Free service
3	Vehicle traffic	0	(not available)
4	Maps	0	About 50,000
5	Airlines	3	Free service
6	Rental cars	0	Free service
7	Hotels	2	Free service

Source: i-mode menu and author's analysis as of early 2003

European cities. However, the evolution of these services in Japan tells us a lot about when and how they will become important to U.S. firms and consumers as location-based technologies diffuse.

CAR NAVIGATION SYSTEMS

Car navigation manufacturers in Japan and to a lesser extent in Europe provide a technology that is difficult for people in other countries like the United States to understand. GPS satellites, detailed maps, sophisticated displays, and high-speed processors provide fast and effective route guidance that is far superior to what can be obtained in the United States with systems like GM's OnStar. Using a device that looks like a TV remote control, users can control a display that is several times larger than a PDA screen. Simply by inputting your destination either as an address or, in the case of commercial businesses, a phone number, or by using the cursor to select a location on the map, the system will within a few seconds identify the optimal route; it then provides voice and visual guidance for the entire trip using data from the GPS (global positioning system) satellites.

All of this comes at a high price of course. CD ROM-based systems cost almost $1000 while DVD- and hard-disk-based systems cost between $2000 and $4000. One reason for the high cost of these systems is that unlike the PC, television, or other well-developed consumer industries in the world, few of the components are standard. Most manufacturers use proprietary CPUs, maps, map engines, operating systems, and displays. Furthermore, they compete by adding new features including better displays, more detailed maps and information, DVD movies, and hard disk drives. Interestingly, the GPS function only adds between 20,000 and 30,000 yen ($160 to $250). Instead, it is the sophisticated display, processing capability, and DVD player. In the latest systems, hard disk drives and TV playback capabilities also add costs.

The Disruptive Nature of the Mobile Internet

The mobile phone and the Internet provide new ways to obtain car navigation data. Instead of downloading map data from expensive CD ROMs or DVDs, it is possible to download the data from a proprietary Internet server using a mobile phone. The first firm to offer this kind of service was Xanavi, which is a newly formed joint venture between

Nissan and Hitachi. Like many providers of disruptive technologies, Hitachi is not the leading provider of existing car navigation systems.

Xanavi's car navigation system is sold as an option on one of Nissan's economy cars called the March. The system is priced at about 50,000 yen ($400) as a built-in system, or one-half the price of the cheapest car navigation system on the market. Of course, in many ways it does not have the capabilities of the existing car navigation systems. It looks like a radio since it is about 8 inches wide, 2 inches tall, and 10 inches deep. The display is 4.1 inches on the diagonal as compared to 7 inches for a normal car navigation system. A bigger display would raise the price not only in terms of the hardware but also in terms of a pop-up display. Pop-up displays are typically used with expensive car navigation systems in order to utilize the dashboard geography more efficiently.

The system contains a rough map of Japan that is stored in a compact flash card. The limited memory capability of the flash card only allows the storage of map that has a scale of one to 10,000. Users download more detailed map information and route details by attaching their mobile phone to the car navigation system. It takes between 60 and 90 seconds for the system to calculate and download route information (with network speeds of 9600 bits a second) as compared to less than a few seconds with DVD systems. Although the user must pay for the mobile phone connection time while the information is being downloaded, assuming that they don't drift off course, only one download is necessary since the GPS system constantly upgrades the car's position without use of the wireless system.

The route information includes a high-level route map that includes landmarks and major intersections, detailed maps for each intersection where a turn is made and the destination area, and voice guidance. Due to the smaller size of the screens and the importance of reducing downloading time, the maps contain far less information than the conventional car navigation systems.

This is a classic disruptive technology because it is possible to roughly predict how the technology will become comparable to conventional car navigation systems. First, improved memory technology will increase the amount of map data that can be saved in the memory flash card. Second, a packet system would eliminate the time to negotiate the server connection, which can be as long as 30 seconds. Xanavi will make its system compatible with the packet system in the W-CDMA third-generation systems once growth begins to occur in these third-generation systems. The wide use of W-CDMA in the rest of the world will probably drive diffusion of these Internet-based car navigation systems also outside of Japan. Third, the higher speeds available with the W-CDMA data services will further reduce the downloading time to a total time of less than 30 seconds. Other improvements in network, handset, and navigation system technology would be required to make the new system as fast as the conventional car navigation system.

The manufacturers of conventional car navigation systems are responding the way firms usually respond to disruptive technologies. They are ignoring the technology while they add new functions to their high-end product. For example, they now offer television reception and 10-gigabyte hard disk drives that enable users to download music, videos, and routes from their PCs to the car navigation systems. Their customers can also use the hard disk drives to update the maps on the car navigation system, but this requires them to visit an auto parts supplier who will do this for them.

I asked one manufacturer why it would not develop a similar low-price system for economy cars. His response went something like this: "Owners of economy cars (in this case the *March*) are housewives and they don't need a car navigation system to go the supermarket. The major consumers of car navigation systems want fast response time and large displays. The new systems will never succeed."

It is too early to know if the incumbent manufacturers are wrong about the market for the Internet-based systems. Only a few percent of March's sales have included the car navigation option. But as long as enough people buy the product to justify Xanavi's continued involvement, improvements will be made, thus causing the market to expand. A critical question is whether these improvements will occur faster than the cost reductions being made to existing CD-based car navigation systems.

Contents for Car Navigation Systems

Of course maps and other route information aren't the only information that can be downloaded with car navigation systems such as the one just described. The Japanese government and manufacturers of automobiles and traditional car navigation systems have been trying to do this for years using a variety of technologies. For example, the Japanese government taxes car navigation systems and uses the proceeds to offer a free information system via a radio connection called VICS (Vehicle Information Control System). The most popular service indicates traffic levels on the car navigation display by highlighting the streets with red (most crowded), orange (medium), and green (not crowded) colors.

Japanese automobile manufacturers also offer proprietary information systems some of which are comparable to GM's OnStar System. For example, Toyota and Nissan have offered voice information services for a monthly fee where the voice information is accessed via a mobile phone. Unfortunately, these services have been far less successful than the GM OnStar program partly because the most important service - the route support service - is already available in the car navigation system.

The future is clearly with open Internet services that use general-purpose technology such as car navigation systems and mobile phones. While once costing several hundred dollars, the hardware needed to make these connections has become standard items on the newest car navigation systems. Users can now download more than 20 different i-mode contents, and most of them include longitude and latitude data. Thus, when a user accesses (for example) a restaurant on a restaurant site, the location of the restaurant is automatically displayed on the car navigation display. And adding this longitude and latitude to a firm's content site cost less than $10.

Unfortunately, just as in the mobile Internet, user learning is a necessary condition for creating a critical mass of users in this market. Most users don't think about searching for information on their car navigation systems because they have never done so. And because few people do this, few firms offer contents for the car navigation systems. Mail and entertainment were the initial applications that started this user learning for young people in the mobile Internet, but so far applications that play a similar role with business users have not emerged in car navigation systems.

Information on parking garages is one candidate. As long as the parking garage has implemented an information system that tracks the number vehicles entering and leaving the site, it can provide this information on a web site. If the parking garage tracks the status of each space, then it can also take reservations, something that some people are probably willing to pay extra for. Park 24 already offers such a service in Japan, but so far there have been very few users.

Park 24 and other content providers expect the market to grow once voice services are possible. This capability is of course what makes the GM OnStar system popular. Pushing buttons and looking at screens is a dangerous activity while driving. And adding such capabilities is much easier with car navigation systems than with phones due to their larger processing capability. Car navigation systems already provide voice-based directions, and single-word voice recognition systems are already possible with phone. It is expected that voice-based markup languages will enable Internet contents to be made accessible on these car navigation systems by 2004.

It's not clear whether the traditional or Internet-based car navigation systems will be more widely used to download contents. The traditional systems offer larger displays while the Internet-based systems are likely to have better network functions, and their users may be more inclined to use these network functions. In any case, it is possible that people will eventually access information about destinations such as parking lots, restaurants, and stores along with other information like news and stock information. Thus, car navigation systems can be seen as a new means of Internet access that will probably evolve in ways that are different from both the mobile Internet and the PC Internet.

TRAIN AND BUS INFORMATION SERVICES

Train and bus information services have little meaning for car navigation users and typical automobile users. But they have a great deal of meaning for "personal navigation," an activity that many people engage in large cities. They help answer the question of "how to get there." Like the Internet-based car navigation services, the mobile phone as a personal navigation system is a disruptive technology when compared to both the car navigation systems and the PC Internet. As discussed above, this has caused train and bus and destination-information services to be far more successful than map services, the latter of which was expected to form the basis of mobile Internet navigation services since they had been so successful on the car navigation systems.

PC software that calculates the optimal set of trains was first offered in the early 1990s, and in the late 1990s other firms like Toshiba and Japan Railways (JR) began offering train information services first on the PC Internet and later on the mobile Internet. The basic functions offered in these various products are the same for each medium. Users input the departure and destination stations, and the service computes the optimal set of trains in terms of time or fares. Users can also input desired departure or arrival times, and the services will display departure, transfer, and arrival times based on actual timetables. If users input bus stops instead of train stations, the services will display the appropriate buses, times for the buses, and, if appropriate, the best set of intermediate trains.

The simple nature of this service makes it easy to offer on the mobile

Internet. Furthermore, many mobile sites have made it possible for users to register up to a certain number of stations so they do not have to input them each time they access a timetable or a route selection with the station as a departure or destination station. This reduces the number of inputs necessary to obtain timetables or route selections. Other train and bus sites suggest previously input stations as candidates when a user begins a search.

These simplifications and the simple nature of the service mean that users can easily input this information and the output is also easy to understand. This lack of disruptiveness has also caused there to be a strong overlap in users where many of them use both services. Many people use the PC service during the day, with peaks at noon and about 5 P.M. before they leave the office, and the mobile services at night. Interestingly, the peak for mobile services is near midnight, when people are checking to see if they have time for "one last beer."

In the United States, the bus information services are clearly a potentially larger application than train information services. In Japan, providers of the train and bus information services expect the bus services to become more popular as GPS or other position services enable the firms to offer real-time estimates on bus arrival and departure times. Currently, unlike the trains, bus arrival times vary widely due to changes in vehicle traffic, and thus the timetables are not very useful. By the end of June 2003 more than 40 bus companies or about 10% of Japan's bus companies were offering estimates of bus arrival times on their mobile Internet sites. For example, Tokyo City Bus started such services in January 2003 and its site was receiving about 100,000 accesses a day in June 2003. Its main reason for offering such a service is to slow and hopefully reverse the decline in bus riders that has been occurring in its service (and elsewhere). Tokyo City Bus only had half the number of riders in 2002 as compared to the peak of 4.6 billion in 1998.

The basic business model for the train and bus information sites in Japan is a monthly charge of about 100 yen per month. The leaders are the first entrants, Toshiba and Japan Railways (JR). JR has more traffic than Toshiba since it offers a larger amount of functions for free, and it has a higher brand image than Toshiba in this industry. But Toshiba still has the largest number of paying subscribers due to its earlier start and greater functionality in the areas mentioned above and also in maps and destination information. As discussed later, Toshiba is offering this greater functionality in order to compete against not only JR but also other firms who are trying to create more sophisticated business models that rely on reservations and not monthly fees.

DESTINATION INFORMATION SERVICES

Destination information services address the issue of "where do we want to go?" Magazines have provided this type of information on restaurants, movie theaters, concert halls, amusement parks, hotels, and parking lots for many years. In the United States, it is often local magazines that play this role, while in Japan it is national magazines like Walkers that dominate this market. Walkers sells five million magazines per month of which the *Tokyo Walker* is published every week while seven regional magazines are published bi-weekly. Each of these magazines provides information on many of the above-mentioned destinations and other topics. Other magazines or pamphlets tend to specialize on concert tickets, restaurants, or travel.

Many of these firms find it easy to offer both PC and mobile Internet services and reach a different set of customers and needs with each service. Like the train services, it is relatively easy for users to search for and interpret this kind of information even on a small mobile screen. They offer Internet services since these services offer a less expensive way to reach customers than with print media. It is cheaper to download information over the Internet than to print and distribute magazines. It is also cheaper for content providers to collect and manage information over the Internet than over the phone, by facsimile, or in person.

Walkers Magazine

The Walkers magazine offers similar types of information in its magazines, PC site, and mobile site. The PC site is free to users so it charges fees to firms like restaurants to have information about their product or service loaded on the site. It offers two mobile sites each of which costs 250 yen ($2.08) per month. *My Walker* is organized by topic while the *Daily Walker* is organized by day, thus making the latter useful to those looking for information specific to a particular day. Both of these sites have more than enough subscribers to break even. The most popular information concerns restaurants followed by seasonal information (ski resorts in the winter and foliage in the fall) and information on movies/ movie theaters.

In spite of the similar information offered in the magazine, PC site, and mobile site, it appears that each type of media appeals to a different set of users since the peak access times are different. For example, the peak access time for information on 2002 New Year's weekend events varied significantly. The magazines had their peak sales about a month before New Year's Day when the December issue first came out. The PC site experienced its peak accesses on the Wednesday before the New Year's weekend (Saturday through Tuesday), and the mobile site experienced its peak number of accesses on each of the actual weekend days.

These data suggest that the mobile site can potentially complement the PC site and magazine. There are always changes in event schedules, so the PC site and (even more so) mobile site can provide this information and hopefully people will pay for the information. For example, it is hard to get the latest movie information when the magazine is being published, so the mobile site can provide more recent information on movies.

The problem is that the new revenues from the PC and mobile sites currently do not compensate for the drop in magazine sales. Educating people about the Internet and the value of real-time information is a long, slow process. Charging fees on the mobile Internet makes the mobile Internet closer to the magazine than the PC Internet in terms of business models. But how to convince users to pay for this service when they can get it for free on the PC Internet is a difficult proposition. Furthermore, it is organizationally difficult to provide PC and mobile services when historically magazine staffs have been focused on just getting the magazine out the door. Now they need to continuously update that information on the PC and mobile site in order for user to receive the full benefits from the real-time capability of the Internet.

Guru Navi

Guru Navi is the leading supplier of restaurant information on the Japanese PC and mobile Internet. It was established in 1996 and started PC Internet services the same year and began mobile Internet services in 1999, both of which are free to users. It broke even in 2001, made a small profit in 2002, and expects continued profit growth in 2003, in spite of the continued Japanese recession. It had about 1.5 million page views per day in late 2002, of which about 250,000 page views were from the mobile phone.

Users can search for restaurants in terms of area, food type, and budget, and each restaurant contains the telephone number, address, hours of operation, typical prices, the nearest train station and exit, a map, and other information. All of this information is available on both the

PC and mobile sites, and the mobile sites are just another way to provide this restaurant information to restaurant goers. Since almost all restaurants request both services, Guru Navi provides them both as a single package.

Of course, the smaller screens on the mobile phone restrict the amount of information that can be provided in the mobile services. Furthermore, certain policies enacted by service providers like NTT DoCoMo and by map providers have led to other differences. For example, NTT DoCoMo does not allow third parties to provide information (e.g., restaurants cannot provide daily updates on their menu) on an i-mode official site, and the official i-mode map sites do not provide Guru Navi with all their maps for free. The map providers provide only some of the maps for free to Guru Navi users as an advertisement for their paid map services. But only the paying subscribers to the map services can do things like change the map scale, scroll on the map, or attach the map to mail messages (see below).

Initially, Guru Navi focused on increasing the number of restaurants in order to solve the network effects problem. It charged restaurants about 6000 yen ($50) per month to load information about them on the PC and mobile sites, and this fee included reservation and bulletin board service. By the end of 2001, it was able to attract about 21,000 restaurants of which about 45% are in Tokyo.

Beginning in 2002, Guru Navi began focusing on increasing the sales for each restaurant. It offered more discount coupon services, pictures (up to 300 per restaurant), fast updating services for menus (many restaurants update the pages themselves), more detailed explanations, search services for discount coupons, and other services. Discount coupons are mostly offered on low traffic restaurant days like Mondays and on days before the monthly payday. It sends discount coupons to small groups of people (500 - 1000) who have registered for them in a designated area. About 200,000 people had registered for these services as of late 2002. These users can print out the coupons from their PC or show them on their mobile phone screen. The increased number of customers from such services has caused some restaurants to pay Guru Navi between 500,000 and 1,000,000 yen ($4200 to $8400) a year.

These success stories have also enabled Guru Navi to increase the number of restaurants that pay for these premium services. According to Guru Navi, the restaurants that pay for these premium services report higher satisfaction than those who do not use the premium services. Fifty-nine percent of the purchasers of the extensive services claim that they are "very satisfied," while only 32% of the purchasers of the basic services claim that they are "very satisfied."

Guru Navi's largest customers are small chains of less than 10 restaurants. These small chains often use Guru Navi's service when they are expanding the number of their restaurants. Guru Navi believes that the market for such small chains is large since these small chains currently spend about 2 - 3% of their sales on advertisements versus about 5% for large chains such as McDonald's.

MAPS

Maps, like train and bus information services, address the issue of "how do we get there." Unlike train and bus information services, they address this issue at a much higher level of detail. Internet-based maps offer a lot of advantages over printed maps, particularly in terms of the ability to easily change the scale of the map. PC Internet services have traditionally relied on income from firms who pay to have the location of their services or retail outlets loaded on the map.

The primary advantage of mobile Internet maps is that they can be accessed from outside your home or office. Several firms began offering mobile map services in late 1999 and early 2000. As with the PC services, users can search for the appropriate maps by inputting the relevant address, train station, or landmark. So-called "free-word" searches that recognize partially input landmark or other names are critical since with mobile services it is difficult to input and often to remember the full names of buildings, companies, or other places.

These map providers offer simple maps for free and more detailed maps plus advanced functions such as the ability to adjust map resolution (about five levels), scrolling in eight directions, sending maps in mail messages and to facsimile machines, and map saving for between 200 and 300 yen ($1.67 and $2.50) per month. The more detailed maps show train stations, public offices, hotels, facilities, restaurants, and stores. The most popular function is sending maps in mail messages; by clicking on the appropriate icon, a URL for the map is created and included in the mail message.

Unfortunately, the map services probably have less than 10% the number of paid subscribers and 1% the traffic of the train and bus information services. Of course, they charge more than the train and bus information services so they are probably above the breakeven point of three million yen ($25,000) per month. The reason why the map services have low traffic is due to the small screens and their poor resolution. It is very difficult to interpret maps even with the advanced color screens that offer 65,000 different colors.

Disbelievers should cut a hole in a piece of paper that is the size of mobile phone screens (the largest in Japan are 2.4 inches on the diagonal) and place the paper over any paper-based map. You will find that very few street names are displayed on the screen for the simple reason that street names are only shown on certain areas of a printed map. While it is quite simple to find the street and follow its course on a printed map, it is very difficult to download 10 - 20 small mobile maps to find all the relevant street names.

And the difficulty of using maps in the Japanese mobile Internet is in spite of the fact that Japanese use a hierarchy of area addresses as opposed to street names. This hierarchy of area addresses should make it easier for users to understand a mobile map. The problem is that displaying these addresses covers up landmarks, and the partial display of addresses (similar problems exist with street names) can change the meaning of the area. Thus, map providers do not display area addresses and instead rely on landmarks (e.g., large buildings, parks) to help the user find his or her way to their destination.

Landmarks present a new set of problems. The map providers had to reduce the number of landmarks in order to account for the lower resolution of the mobile phone displays. Manually, this would take too long so the firms wrote algorithms that attached priorities to landmarks and chose the landmarks to be included on the maps such that the landmarks would not bump into each other. This has caused the most well known landmarks in some areas to be eliminated. Although improved display resolution particularly from the color displays enables better maps, it also requires firms to continually upgrade the choice of landmarks.

Route services, which are the main application on car navigation systems, are also much more difficult to provide on the mobile Internet due to the lower amount processing capability in mobile phones and the greater capability demanded by pedestrians. Pedestrians have much higher demands than automobiles since there are more choices for pedestrians than automobiles. Pedestrians want to know about whether they should pass through a building or go underground in addition to the best street to take.

B2C INTEGRATED SERVICES

The Japanese content providers are currently trying to integrate map, destination, and train and bus information services. The map providers are trying to offer destination information on restaurants, bars, and stores. The destination information providers such as Guru Navi offer map services while some of the train and bus information services are trying to offer both destination information and map services.

The map services are probably the easiest ones to provide. Many firms have map data, and they are willing to sell these data for a price that covers the marginal cost of modifying the data for the mobile Internet. Even the map providers, who may purchase or own their own map data, are giving away their maps for free to the restaurant services in order to advertise their services. As discussed in the last section, the challenge is to provide maps that overcome the small phone displays.

The destination information services are much more difficult to provide due to the network effects in the industry. It took Guru Navi almost five years to develop partnerships with 20,000 restaurants, and thus this represents a valuable resource. Lists of restaurants or so-called "yellow-page services" are believed to be of little value since they merely include addresses and phone numbers, which is a small part of the service that Guru Navi offers.

Within the destination, train, and bus information services, Guru Navi, NaviTime, and Toshiba appear to be the best positioned. Guru Navi is well positioned in restaurants and providing maps for these restaurants. NaviTime offers the most sophisticated algorithms and most efficient route guidance for trains and buses; it entered much later than Toshiba but is slowly catching up.

Toshiba can use its success in train and bus information services to expand into destination information around train stations. These destination sites include restaurants, bars (which raises the possibility of the streaming services mentioned in Chapter 3), movie theaters, and even parking lots. A focus on the destinations around train stations reduces the size of the network effects problem. Toshiba hopes to collect charges from restaurants, bars, and stores to have information about these places located on Toshiba's site. Toshiba is also trying to combine these destination information services with a free magazine called milKL in much the same way that some mobile shopping sites are combining their sites with magazines. If users input a four-digit code into the site (the four digit-code is listed in the area magazine), the user can directly access the restaurant, bar, or store without searching for it on the menu.

The other train and bus information sites are expanding their services in other ways. For example, JR offers information on the best places to see the fall foliage, hot spring resorts, beaches, and ski resorts in an effort to promote train usage particularly on weekends when usage is low. As discussed in the next chapter, JR also offers smart cards as train tickets and it hopes these cards will be used as forms of payment by stores near train stations. Other train lines offer information services in combination with train passes. By registering their mail address with their train pass, registrants can receive mail alerts about local restaurants, bars, and stores, including discount coupons after they pass through a ticket wicket with their train pass.

B2C AND B2B INTEGRATED SERVICES

It is also possible to integrate B2C services with B2B services. This section looks at the application of this concept to the food and beverage industry while a subsequent section looks at this issue in a slightly broader way focusing on the integration of GPS or other position services with B2C and B2B services.

Figure 7.2 summarizes existing and possible B2C and B2B services in the food and beverage industry. Food and beverage producers have long used EDI to communicate with their distributors and are now moving toward PC Internet-based systems (see number 1 in Figure 7.2). Distributors of food and beverages are trying to convince restaurants, bars, and hotels to order their food and beverages over the Internet as opposed to ordering by telephone or facsimile in order to reduce costs (number 2).The United States is probably ahead of Japan in both these areas, although diffusion is probably occurring slowly also in the United States in the second application for the reasons discussed below. In Japan, Guru Navi and other firms already provide restaurant information to consumers (number 3).

The mobile Internet promotes B2B services partly through their integration with B2C services in at least three ways. First, many restaurants and bars, even those in hotels, often do not have access to a PC in Japan or even the United States. They are too expensive and large for many restaurants and even more so for bars. This is why a croquette restaurant chain called CoroChan introduced an ordering system that is based on i-mode phones into its 550 shops. Its shops average 5.7 square meters (or about 55 square feet). As opposed to sending facsimiles to headquarters, which forwards them to suppliers, stores now place their orders on their mobile phones. This reduces telecommunications costs and missed translations; the latter occurs when headquarters personnel reenter the order into their computer system in order to send the order to suppliers. Furthermore, the new system facilitates the creation of sales reports and also enables the input of time card and

FIGURE 7.2. Potential B2B and B2C Internet Services in the Food and
          Beverage Industry

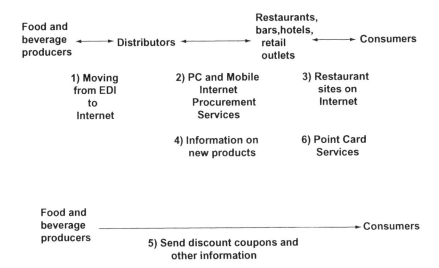

other data via the mobile phones[3].

Other bars and restaurants may require additional incentives to place their orders over the Internet, which brings us to a second advantage of the mobile Internet. Restaurants and (in particular) bars are interested in the latest information on beverages and foods partly since new alcoholic beverages are first consumed in bars as opposed to homes. Distributors may find it easier to convince bars and restaurants to place orders over the Internet if they provide bars and restaurants with more information on new alcoholic beverages (see number 4).

Bars are also interested in what customers enjoy drinking as are manufacturers, who would like to establish contacts with the people who actually consume the food and beverages (see number 5). And since most new foods and beverages are consumed in bars and restaurants long before they are consumed in homes, most consumers think about these things more in bars and restaurants where they have their mobile phones than in homes where they have their PCs. The problem is that manufacturers of food and beverages do not have these contacts while Guru Navi does, which suggests possible alliances between food manufacturers and firms such as Guru Navi. On the other hand, distributors would like to make these contacts with the customers in order to increase their leverage with the food and beverage manufacturers.

Third, as was discussed in Chapter 5, bar codes and the infrared function in the latest NTT DoCoMo phones enable restaurants and bars to use phones as point cards. This provides them with a way to establish stronger relationships with their customers and possibly help the food and beverage manufacturers to make contact with their customers. Distributors, manufacturers, and restaurant sites such as Guru Navi (see number 6) could provide point card ASP services to restaurants and bars that are not large enough to implement the systems on their own.

The concept of the "mobile phone as the new contact point for customer" plays an important role in this new competition. Everybody wants to control access to their immediate customers and, if possible, the final customer. As the mobile phone becomes the new contact point for customers, particularly those young customers who frequent many bars and restaurants, the strength of a firm's portal on the mobile Internet becomes important. The exact combination of services needed to make this happen is still uncertain, and various alliances will likely emerge.

BUSINESS SERVICES

Business users will probably want more sophisticated navigation services than those described in previous sectiond. One possibility is to combine the types of navigation services discussed in previous sections with scheduling software. As someone creates their schedule on their mobile phone or PDA, it should be possible to register maps for destinations and reservations for rental cars and hotels. The maps could be for offices, friend's homes, or restaurants, and each of these maps could be linked to various databases, including their own personal database. One element of the personal database could be business cards, which may be exchanged through infrared linkages between phones. The business cards could provide links to maps, client-related data, and company home pages.

Reservations could also be integrated with these schedules. American's lead in Internet airline, hotel, and car rental reservations probably makes this a better application for the United States than for

Japan. The Internet is used for less than 15% of airline reservations in Japan a few percent of which are done on the mobile Internet. Even if they do their reservations on the PC, business people could save reservation information on their phones via a mail message from the PC site. In the long run the phone could be used as an airline ticket (see Chapter 8).

Providers of scheduling software would be a natural provider of these integrated services. Scheduling software is already available on mobile phones in Japan and is apparently the most popular function on PDAs in both Japan and the United States. PDAs are definitely a better device for these services than mobile phones due to their superior user interface and wide use among business users. The major challenges for PDAs include the small number of people who use their PDAs to access the Internet and the small number of contents that are formatted for the PDA, which are clearly related problems.

More advanced systems would provide ways to update the schedule as new information is received. For example, the scheduler could provide a list of changes to be made and various options if mail arrives from the airline indicating a delayed or canceled flight. Similar things should be possible when there are changes in meetings, weather, or traffic.

While some of these features might first be offered at the domestic level, some Japanese firms are already thinking about these services at the global level. Jordan, a provider of train and bus information services, wants to offer domestic and global travel services to Japanese companies and consumers. It offers its train and bus information services for free in order to develop a large user base, which it hopes it can use to become the leading provider of travel reservation services. Jordan created alliances with Galileo and W-cities in late 2001/early 2002; Galileo was spun off from United Airlines just as Sabre was spun off from American Airlines. W-cities is an English company that is building worldwide maps.

GPS AND LOCATION-BASED SERVICES

We now reach the topic from which most discussions of mobile navigation services start, indeed from which many discussions of the mobile Internet once started! Location-based services are important in the mobile Internet, but they pale in comparison to the importance of portability. As we have seen in previous chapters, the portability of the mobile phone has enabled a completely new set of Internet services to emerge such as ringing tones, screen savers, Java games, simple news, discount coupons, point card services, and others.

In this chapter we have seen how the ability to access and interpret information on the mobile phone has determined the relative success of train and bus, destination, information, and map services. It is easier to download train and restaurant information than to download map information, so the former two have become successful services while the map services have not. And since the former two types of services have also already introduced map services, it is likely that they will be just as capable of introducing GPS services as the map providers will be.

Positioning Technologies

There are a variety of ways to obtain position information each of which provide different levels of position accuracy[4]. Most of these techniques use either the service provider's network of base stations or GPS-compatible devices to identify an object's position. The network of base stations can provide information on the closest base station to a phone or a more accurate position using triangulation between multiple base stations. In a downtown area this can theoretically provide an accuracy of about 300 meters.

NTT DoCoMo introduced a service called i-area in 2002 that uses the closest base station to determine the phone's location (J-Phone offers a similar system). By clicking on the i-area service and one of its compatible contents, users can more quickly find information about the local weather or the closet restaurant or movie theater. In reality, the low accuracy of the positioning technology meant that the user might be saved one or at the most two clicks on their phone. More sophisticated techniques like triangulation between base stations can provide additional accuracy albeit at the expense of power consumption.

GPS is based on 24 satellites and 5 monitoring stations around the world that enable the satellites to broadcast a signal that can be used as a reference in determining the position of a GPS receiver. In combination with gyroscopes and under the right conditions - for example, when line of sight is possible - accuracies as high as several meters are possible. Unfortunately, phones do not contain gyroscopes and ideal conditions often do not exist; thus the accuracy is often no better than 100 meters.

The biggest problem for using GPS in phones is power consumption. Although this is not a problem in vehicles where GPS receivers are currently widely used (i.e., conventional GPS systems), high power consumption precludes the use of pure GPS with phones and thus requires some form of "assisted GPS." One form of assisted GPS involves placing chips in phones that estimate the probable location of the satellites based on their expected paths and the location of the closest base station. One problem with this approach is that many services providers are constantly adding new base stations (particularly in the case of 3G services) and the number of these base stations can be very large in some cases (e.g., all GSM base stations in the world). The latter problem can be solved by improvements in phone memory and processing capability, and we will return to this later.

A second and more popular method is network-assisted GPS, which was introduced by KDDI in 2002. With network-assisted GPS, most of the GPS calculations are done at the server level, thus reducing power consumption but driving up communication costs. KDDI introduced the first GPS-enabled phone in early 2002 and it is expected that most new KDDI and other Japanese phones will have similar capabilities by late 2003 and 2004, respectively. KDDI's system combines network-assisted GPS technology from Snap Track (a subsidiary of Qualcomm) with base station triangulation technology from Qualcomm and provides an accuracy of almost 10 meters. The addition of the base station triangulation function enables the system to also provide location information even when users are not in clear sight of a GPS satellite (e.g., near or in buildings), which is typically a problem with systems that rely only on GPS.

At current packet charges of 0.2 yen per packet (128 bytes per packet), a single position request costs almost 10 yen using KDDI's network-assisted GPS technology in early 2003. If you were to update a phone's location every 5 minutes for one eight-hour shift, you would incur almost 10,000 yen ($84) in operating charges as compared to almost zero with conventional GPS systems. This experience suggests that network-assisted GPS will not be used in car-based applications until the packages charges have dropped dramatically.

Network-assisted GPS is also time-consuming, which will also slow the adoption of this technique. It takes about 45 seconds to download a map containing your present location, including 15 seconds to download the GPS data, 15 seconds to send these data to the server and carry out the calculations, and 15 seconds to download the map. Of course, reductions in packet charges and faster network speeds, like those discussed in Chapter 2, could reduce these communications costs and times significantly.

A combination of the network-assisted and pure GPS approach might be feasible when doing continuous updates in a short time frame. The network could handle the initial calculations of probable satellite locations, and the phone could search in the same vicinity in the position updates. If the user is not moving very fast, it might be possible to provide a number of updates without any communication with the phone network.

KDDI claims that one of its phones released in October 2003 has such capabilities plus the ability to provide route guidance via voice. Reportedly it could provide route guidance for a 15 minute walking distance in a few seconds and the guidance would only incur about 20-30 yen in communication costs. Improved application processors should continue to reduce these costs. The next two subsections look at the implications of these GPS services for both consumers and business people.

Consumer Services

Japanese train and bus, destination, and map information providers have all added GPS functions to the map services that they provide for KDDI users. By putting a compass in the phone (available in one phone in 2002), the phone's location appears on the map along with the direction you are facing. The latter is very important and raises the value of maps significantly. As discussed above, it is very difficult to interpret the maps, much less where you are on the map.

Exchanging GPS coordinates with friends can be useful, and several content services that support such activities have already emerged. For example, one of the map providers, Zenrin, offers a service whereby subscribers can track their friends' phones. Simply by inputting your friend's phone number (authorization from the user is of course required), their location appears on your phone's (or PC) display.

Destination services are also integrating their contents with GPS services. As mentioned earlier, it costs less than $10 to add longitude and latitude data to restaurant or other destination's database. Already Toshiba has loaded the data for most train stations in Japan thus allowing people to use their service to also get to the closest train station. Guru Navi has started including longitude and latitude data for restaurants in its database, and map sites are doing similar things.

It should also be possible for some of these sites to offer discount coupon services that are location-based. Train and bus (which might build off the location-based services mentioned earlier), map, and restaurant information services might offer these services. But these location-based discount coupons cannot be Spam; they must be services for which the user registers. And the people who want to receive these location-specific coupons will probably be a certain percentage of those who register for discount coupons rather than a driver of the discount coupon market.

Other types of services can be offered when you combine the PC Internet with the GPS phones. For example, Toshiba offers a consumer service that enables people to track phones on their PC screen. Naturally, the service requires authorization from the phone user. The service costs 1000 yen ($84) to register, 300 yen per month ($12.60), and either 70 yen (from the Internet) or 300 yen (from a call center) for each location request. Secom, Japan's leading Security Company, offers a similar system that is slightly more expensive but includes emergency services. In both services, users can also register via regular mail and receive a facsimile containing a map displaying the phone's location simply by contacting the call center. Parents concerned about their children's safety, people trying to locate their friends in crowded urban areas, dating services, and jealous spouses are some of the consumer applications that are expected to emerge.

For example, many parents want to check to see whether their teenage daughter is staying at Brenda's and not Paul's house, but at the same time they don't want to send a message of distrust to their daughter. By checking the location of the phone on the PC screen, your daughter may not know whether you are checking up on her. Of course, privacy laws may require service providers to notify users when they are being tracked. In this case, some parents may buy their teenager a phone in return for the right to follow their activities.

NTT DoCoMo believes that more than 10% of its subscribers will use these functions since about 10% of its PHS users subscribe to these functions. PHS is a mobile phone system that relies on a dense system of very low-power base stations. By knowing the closest PHS base station to the PHS handset, the system can calculate a user's position to an accuracy of between 100 and 300 meters. Since NTT DoCoMo's main mobile phone system will have higher accuracy and not require special subscriptions, it is possible that more than 10% of NTT DoCoMo's 40 million subscribers will eventually use the GPS services.

In the long run sites may be designed for specific GPS locations. Users may be able to access or input information about specific restaurants, stores, or even a place on a mountain trail. Although many Japanese firms are already developing the technology for such space tags, there are still many unanswered questions about business models and site management.

Business applications currently represent the largest market for these location-based services due to the high costs of accessing position information. Several firms have offered services that are based on PHS and cellular systems with varying degrees of success for the last several years, and recently several firms have begun offering services based on KDDI's GPS system. For example, KDDI, Toshiba, and Secom started ASP services in the fall of 2002 that use KDDI's GPS phones. Like the consumer services discussed above, the business services enable firms to track phones that are owned by their employees on a PC screen. An important function in these ASP services is the ability to manage the tradeoff between position accuracy and packet charges. Users can set the frequency with which the system accesses a phone's position information and thus the relative accuracy and packet charges. For example, by reducing the frequency of measurement from every 5 minutes to every 30 minutes, the cost of tracking a device for one eight-hour shift drops from about 10,000 yen ($83) to about 1500 yen ($12.50).

It appears that the first users of these business services will be quite different from the users of conventional GPS systems. While transportation firms are the largest users of the conventional GPS system, the largest numbers of applications that are being implemented by KDDI's 200 participating firms (as of June 2003, which is up from 130 in January 2003) involve sales, service, and home health care. Many firms want to monitor their sales people to understand who they are visiting and where. Maintenance groups, whether they are independent companies or departments in large companies, would like to use information about current worker location to select the appropriate worker to be assigned to a specific call. Home health care is a growing industry in most countries due to the increasing number of elder people who require medical care in their homes. To reduce costs, many home care workers visit multiple homes in a single day, and health care companies would like to assign the closest person to a specific emergency call.

While many small firms may implement these services independently of their current information systems, large firms will probably want to integrate these services with their ERP (enterprise resource planning), SAP (sales automation planning), or CRM (customer relationship) systems. KDDI has already received many requests to add new functions to its GPS system to handle these other functions. For example, home health care companies want their employees (e.g., nurses) to input work times and travel expenses (e.g., train fees) as they are being sent to various homes.

Another example can be found with trucking companies. GPS functions, whether they are conventional systems or in phones, can be used in combination with mobile mail functions to facilitate the recording of delivery activities. While recording this information can also be done with special handsets, mobile phones are cheaper and everyone owns one. For example, in a system offered by DoCoMo Machine Communications, a subsidiary of NTT DoCoMo, the GPS system (a conventional system) recognizes when a driver reaches a designated point, and at this time the driver receives mail that contains a URL. After clicking on the URL, a form is downloaded on the phone screen where the driver can input information about the delivery including the number and type of items delivered, their status, and the mileage displayed on the odometer. Other information including firm name and time can be automatically included in the form based on a schedule that was previously input into the computer system.

The combination of camera phones and GPS may also be a large application. Many firms offer ASP services where the location of a photo taken with a camera phone is automatically registered with the photograph. This facilitates report preparation for example with road construction. The Japanese Ministry of Transportation plans to use camera phones to record potholes and other road problems in order to speed up such repairs. Environmentalists want to use such services to record environmental problems, for example the illegal disposal of waste.

SUMMARY

The disruptive nature of the mobile Internet has caused an unexpected set of users and applications to dominate the early navigation services. Train and bus services have done better than maps in the non-GPS applications, and workers that commute by train are bigger users of GPS applications than conventional users of vehicle-based GPS. Lower packet charges, greater processing speeds, and new business models will expand the applications for both GPS and non-GPS applications. Some of these new business models will include the integration of B2C, B2B services, and scheduling services.

NOTES

[1] *USA Today* article summarized in *CTIA Daily News*, December 5, 2002.
[2] According to suppliers of these conventional GPS devices, they typically weigh several hundred grams and cost as much as $500.
[3] Source: NTT DoCoMo's home page: www.nttdocomo.com/I-mode/expanding/forbiz.html
[4] For example, see Hjelm, J., *Creating Location Services for the Wireless Web*, NY: John Wiley & Sons, Inc., 2002. This section relies heavily on Chapter 2 from this book.

Chapter 8

Phones as Tickets and Money

Phones represent a new form of money and ticket that will continue to reduce the importance of physical cash and paper tickets in the global economy. The usage of checks began to grow after World War II and by 1984, 58% of America's consumer purchases were made with checks as compared to 36% with cash and 6% with credit cards. Since then, credit usage has exploded and in 1996 Americans made 22% of their purchases with payment cards as opposed to 57% with checks and 21% with cash[1]. While other countries may use debit more than credit cards or use more bank transfers than checks, the move away from physical cash is a global phenomenon with currently cash primarily used in low-denomination purchases.

These trends offer large benefits to both consumers and businesses. Credit cards reduce the need to carry large amounts of money and also provide the user with a short-term loan, while debit cards reduce the need to carry smaller amounts of money. Businesses benefit from increased sales by people who do not have enough cash to make a purchase. Furthermore, both credit and debit cards reduce the risk of having money stolen from both consumers and businesses.

On the other hand, problems with existing credit and debit cards provide opportunities for new forms of money. Both credit and debit cards require rather long processing times in stores due to the authorization functions, and credit cards have rather high transaction fees for stores. The former makes it difficult to use these cards in small purchases and the latter will encourage stores to consider new forms of

TABLE 8.1. Competing Technologies for Phones as Tickets and Money

Technology	Phone Requirements	Speed	Two-Way Communication	Security Issues	Likely First Application
2D Bar codes	Mail function	Slow	No		Mileage Cards
Infrared	Infrared port	Slow	Yes	Directionality exists	Credit card purchases
Non-contact smart cards	Smart card function	Fast	No	Short distance	Transportation

money.

Paper tickets face a slightly different form of problem. They require some form of pick-up or delivery even when they are purchased online. This makes it harder to do same-day purchases; furthermore, online purchases also make it easier for scalpers to dominate the lotteries where people win the right to purchase a ticket. The scalpers annoy consumers when the consumers must pay a higher price for the ticket, and they annoy ticket promoters when the scalpers don't end up exercising their right to buy the tickets leaving many unsold at the last minute.

COMPETING TECHNOLOGIES

Table 8.1 summarizes a number of technologies that can enable phones to be used as tickets or money. The most simple form of technology is a two-dimensional (2D) bar code that can be downloaded from the Internet or received in a mail message and subsequently displayed on the phone screen. Two-dimensional bar codes provide more security than one-dimensional readers, and readers are fairly inexpensive. Another advantage of this technique is that it can be used with any mobile Internet- or SMS-compatible phone.

The problems with 2D bar codes include security concerns and low speeds. It can take more than one second to authenticate them and multiple scans are sometimes required, which is too slow for many applications. They appear to be best suited for loyalty cards and perhaps small purchases where fast processing times are not very important. For the same reasons, using the phone's mail function to send 2D bar codes or other information to cash register or ticket scanning machine is probably also too slow.

Infrared technology provides two-way communication and also high security due to the directionality of the signal. The disadvantages of infrared technology include their slow speeds and the need to put these functions in phones. Although the necessary infrared technology is diffusing rapidly in Japan and Korea, it will still be several years before these phones represent more than 50% of the phones being used in these countries. Europe and the United States will probably be somewhat slower, although a successful mobile Internet service does not appear to be a prerequisite for the use of phones as tickets and money.

Smart cards provide similar or somewhat lower levels of security than infrared technologies but have much faster processing times, particularly if they are non-contact smart cards, and they are available now. Unlike the infrared technologies, firms can build a user base of non-contact smart card users and later transfer them to phones as the non-contact smart card technology is put inside phones. Smart cards are credit-card-size devices that include an electronic chip that makes the card a small computer. GSM phones have contained contact-based smart cards for many year, and credit card companies like Visa and MasterCard have been trying to introduce these contact-based smart cards for years since they believe the cards can reduce transaction costs and the rate of fraud. Unfortunately, the credit card companies have had little success in convincing stores (or even PC users) to introduce the new readers, and the most popular application for smart cards today is transportation.

Transportation applications use non-contact cards since they have much faster processing times (about 0.2 seconds) than contact cards. As the name suggests, non-contact cards can be read without contact between the reader and card, typically at a distance of about 10 centimeters. One reason these non-contact smart cards are so fast is that they are prepaid cards and thus there is no reason to check the user's identity. The disadvantage of these prepaid cards is that they are like money. If they are stolen there is no way to get them back and therefore most of the cards used in Japan are only programmed to hold up to about United States $US 50.

There are three types of these non-contact cards, which are quite unimaginatively called Type A, B, and C cards. Type A cards (based on technology from Philips) are primarily used in telephone card applications while Type B (an improved version of Type A cards) and C cards (based on technology from Sony) are primarily used in transportation applications. More than 30 million of these non-contact smart cards were being used in transportation applications in Asia in late 2002 most of which are based on Sony's technology (Korea uses Type B cards). European subway systems are also a major user of these cards. In Japan, convenience stores and concert ticket providers are also experimenting with Sony's technology due to its fast processing times and success in transportation applications.

Mobile service providers would like to put these non-contact smart card functions in phones in order to provide additional benefits to their customers and to receive some percentage of the proceeds, even if the percentage is small. Of course, intense negotiations are likely to occur over small changes in this percentage, and these negotiations are likely to present bigger barriers to placing the smart card functions in phones than the actual technical problems[2]. The service providers want to use non-contact cards not only because it is difficult to use contact cards in phones for physical reasons but also because the applications for non-contact cards appear to be much larger than those for contact cards.

These non-contact cards are highly disruptive for credit card companies and other financial institutions. Their faster processing times are less relevant for retail outlets and restaurants than for transportation and convenience store applications. And if a widely used standard for non-contact cards were to emerge either by itself or inside of a phone, the cost of readers will decline and stores might introduce readers and accept the cards, particularly if the cards have lower commissions than current credit cards (about 1.8% in the United States and 3% in Japan). This would enable the providers of these non-contact cards, or the phones they are in, to expand their market from transportation and concert tickets and convenience stores to higher denominational purchases where credit card companies make their money.

However, the battle is just starting and it involves a number of different players. Some firms are promoting specific types of smart cards while service providers and phone manufacturers can make decisions about which type of card to put in the phones. As with most standard battles, firms that generate the initial applications often win due to the importance of network effects. This chapter first looks at train applications followed by entertainment tickets and money. It concludes with a brief discussion of biometrics, other applications, and the possible end of cash.

TRANSPORTATION TICKETS AND RESERVATIONS

Many people expect smart cards to replace most train and bus tickets, particularly those that are read automatically with magnetic readers, over the next few years. Automatic ticket reading machines are widely used in subway systems and in large train stations particularly in Asia. Riders insert a paper ticket that has a magnetic stripe embedded in the ticket into the machine. The machine checks the ticket's validity and allows riders with valid tickets to pass through a gate. The use of these machines can decreases fraud and the costs of personnel, particularly when tickets are bought from automatic ticket machines. Of course, the use of these machines assumes sufficient volumes to justify their implementation.

Many of these Asian train and bus lines have also offered prepaid magnetic tickets since they enable users to avoid long lines for ticket machines and in some cases receive a discount. Discounts are particularly common with so-called commuter passes where the first and last stations are designated on the pass. They are less common with general-purpose passes that can be used with any combination of train stations in spite of the fact that some people lose the passes before they have used the card's full value. In any case, prepaid cards are also beneficial to train companies since they reduce ticket dispensing machine costs including maintenance and money handling costs. Handling bags full of change and the machines that handle these coins represent several percent of a train company's operating costs.

Smart Cards

The next stage in user convenience and lower costs for train and bus companies has been the implementation of smart cards. These cards make it easier to combine commuter and general-purpose passes and they make recharging possible, both of which are convenient for users and reduce distribution costs for the train companies. Users can recharge the smart cards at ticket machines and when the smart cards are embedded in phones, users will be able to recharge the cards online by downloading money from bank accounts. Recharging cards is a particularly big problem in places where bus riders do not visit large train or bus stations. Placing the smart card inside a phone eliminate the need for these riders to visit large stations or for train and bus lines to install such machines.

Since they are non-contact, smart cards also reduce the implementation and maintenance costs of ticket reading machines. The lower implementation costs will make it possible for even small train stations to implement automatic ticket reading machines. This will further reduce personnel costs and increase the possibility that the same card can be used with multiple train and even bus lines, which will be a large benefit to users. There are more than five million users of these cards in Tokyo and other regions in Japan are in the process of implementing them (Osaka started in October 2003). The emergence of a standard for non-contact smart cards would provide further incentives for train and bus lines to use a common smart card in Japan; this has already happened in Korea and Hong Kong. It is possible to use one card for trains and buses in Seoul and Hong Kong and in Hong Kong, the same card can be used for purchases in convenience stores.

The increased convenience and lower costs from smart cards may help train and bus companies reverse the decline in riders that continues to occur as rising incomes increase car ownership. Japanese train and bus companies have been fighting to retain riders for years, and similar trends will likely soon begin to occur in countries like China. This makes smart cards (and the GPS services discussed in Chapter 7) important tools in their struggle for survival and in the struggle for a better environment. It is well known that the use of public transportation reduces car traffic and pollution and technologies that can increase the use of public transportation are environmentally beneficial.

Internet Reservations and Smart Cards

Enabling consumers to reserve train tickets on the Internet can also increase user convenience and using phones as tickets supports these Internet reservations. A relatively small Japanese train company, Odakyu, was the first firm in the world to use phones as train tickets beginning in July 2001. Odakyu riders can make reservations on the PC and mobile Internet and use their phones as tickets on Odakyu's express trains. After the users reserve a seat either on the PC or their mobile phone (the latter of which requires about six screens of operations), they receive a ticket via electronic mail. The ticket price, which can run as high as 2000 yen ($16.67), is automatically discounted from a prepaid account.

Riders do need to purchase and use a non-reserved ticket to pass through the ticket wicket. But they are rarely asked to show their reserved tickets on the train to conductors since conductors download data on ticket reservations to a PDA and check to see whether any occupied seat has not been reserved. It is only in this case or when riders cannot agree on the ownership of a seat that conductors ask riders to show their reserved ticket that is stored in the phone. It is also possible to automatically check these bar codes using scanners as Jeansmate does with discount coupons (Chapter 5).

Interest in Odakyu's service grew rapidly so that by mid-September 2001, 10 - 20% of the seats on evening trains and 13% of the seats on morning trains were being reserved using mobile phones. However, it appears that only a few Japanese train lines will adopt Odakyu's approach due to differences in reserved ticket prices. Odakyu's approach is useful when the price of the reserved ticket is low, and thus the chance of fraud and the desire to use credit cards is low.

As the price of the ticket gets more expensive and demand for credit card services increase, higher security is needed. For example, another small Japanese train company, called Kintetsu, began taking mobile and PC Internet reservations in March 2001. Instead of using their phones as a ticket, users receive their ticket either from a ticket window or from a kiosk by inputting their credit card into the kiosk after they have made a reservation on the PC or mobile Internet.

Japan Railways (JR) implemented a similar system in its bullet train lines (Shinkansen) in September 2001; the key difference is that users must first become members of the service and receive a member's card. This is due to the high costs of the tickets (typically over $100) and thus the high chance of fraud. After riders make a reservation on their mobile phones or PC, they insert their member's card into a special machine at the station and receive a ticket. Several hundred thousand people have already registered for the service, and hundreds of people can be seen using the machines to receive their ticket on any late afternoon in Tokyo station.

Several factors have been preventing a wholesale shift to Internet reservations in Japan. One is that discount tickets (available in so-called discount shops) are not yet available to riders who reserve tickets on the Internet. JR is planning to change this when it makes other changes to its ticket pricing system in the fall of 2003. Interestingly, it expects that these price reductions will actually increase its margins since it will be able to sell more tickets directly to the final users as opposed to selling through the discount ticket shops.

A second problem is riders' inconvenience of picking up tickets from a kiosk when they are in a hurry to make a train, particularly an early morning train. Not knowing how long it will take to get your ticket or if there will be problems with getting the ticket requires riders to arrive early at the train station. Smart cards will solve this problem because they enable the use of the smart card as a ticket. When the smart card is used to pass through the appropriate ticket wicket, the system merely compares the reservation and member information. Like the Odakyu case described above, the conductor merely checks to see whether any occupied seat has not been reserved.

The ability to make reservations over the mobile Internet will have a dramatic effect on the world's train systems. Most train stations have long lines at their ticket windows, particularly in the late afternoons when people are on their way home. The ability to make or change their ticket reservations from their phones will enable riders to make these changes while they are on their way to the main train station and thus avoid the wait in line. An additional benefit is lower personnel costs for train companies.

Furthermore, mobile Internet reservations may also increase reserved seat utilization since some people initially reserve seats and then end up riding in the non-reserved section of trains. On the Japanese bullet trains, this is because it is typically much faster to line up in front of one of cars that contain non-reserved seats as opposed to waiting to have their reservation changed and for the train to actually depart. In these cases, the person is actually using two seats since Japan Railways does not know that the person has boarded another train and thus cannot give his or her reserved seat to someone else. Mobile phone reservation changes may decrease the numbers of these "two-seat users" and thus allow JR to offer a fewer number of bullet train cars.

CONCERT AND OTHER ENTERTAINMENT TICKETS

Many people also expect smart cards or phones that contain infrared technologies or smart card capabilities to replace most concert, sports, and other entertainment tickets over the next few years in Japan and Korea and subsequently in other countries. Like the transportation tickets, these technologies are the next step in a series of improvements that have been made in ticket selling. These include telephone and Internet reservations of which the latter includes both PC and Internet services.

Japanese firms are focusing more on these technologies for concert tickets than for other forms of tickets due to the larger amount of Internet (particularly mobile Internet) reservations with these than with other tickets. Regular concertgoers tend to be young, somewhat affluent, and big users of the Internet. Many seats for baseball and other sporting events are sold as season tickets, and it is perceived that the purchasers of the non-season tickets are not big users of the Internet. This is also probably true in the United States where white-collar workers are more likely to use the Internet than blue-collar workers.

The Internet is quickly becoming the preferred method of buying concert tickets. Although Japan was slower to begin selling tickets over the PC Internet than the United States, sales over the Internet in Japan have quickly accelerated over the last few years, particularly through the success of Japan's mobile Internet services. In February 1999, Lawson and Pia began offering i-mode concert ticket services and E-Plus, a partly owned subsidiary of Sony, started these services in April 2000. Mobile Internet services have enabled these Japanese firms to extend their Internet sales, which brings lower transaction costs, to a new set of customers.

Table 8.2 summarizes the percentage of tickets sold over the Internet by Japan's largest ticket sellers as of mid-2002. While there have been some increases in Internet purchases since then, Pia and Lawson still reserve most of their tickets via telephones, and they either deliver the tickets by mail or have them picked up in their many ticket outlets. Pia's ticket outlets represent a large physical investment that may be rendered useless by Internet reservations, and Lawson uses its convenience stores as ticket outlets. Users input their reservation number into special terminals that Lawson has installed in its convenience stores and receive a ticket, which they then pay for at the register. E-Plus does not have ticket outlets and thus delivers the tickets by mail and primarily depends on Internet reservations.

The Purchasing Process

In order to understand the need for smart cards, we need to look at the

TABLE 8.2. Ticket Shares by Seller and Method

	Pia	Lawson	E-Plus
Total share of tickets	#1	#2	#3
Percent of tickets sold over the Internet	6%	10%	85%
PC	3%	5%	75%
Mobile	3%	5%	10%

Source: Interviews and author estimates

purchasing process for concert tickets, which appears to be similar in most advanced countries. Concert tickets that are ordered by telephone or online are either delivered to the buyer or picked up at a specific location by the buyers. If the buyer purchases the ticket well in advance of the event and pays with a credit card or a bank transfer, it is simple to deliver the ticket by mail to the buyer's home or office. If the buyer wants to purchase with cash or if they are not purchasing the ticket well in advance of the event, the buyer must pick up the ticket from a specific location.

Rock concerts have special characteristics. First, many high school and college students do not have credit cards or bank accounts, and they often do not want to ask their parents to purchase the tickets for them. Thus, they must pay for and pick up the tickets at a specific location. In Japan, this provides firms like Pia and Lawson with a strong advantage since they have many ticket outlets. But as more young people acquire credit cards and/or bank accounts or smart cards become more widely used, this advantage can become a disadvantage as is described below.

Second, many rock concerts sell out very quickly. Many readers probably remember television images (or experienced it themselves) of hundreds of young people lined up for hours to get good seats. Ordering by telephone only helped a little because many people were forced to spend many hours waiting on the telephone.

The PC Internet and mobile Internet have partially solved this problem and created a set of new problems. Japan's Internet ticket sellers enable people to register for lotteries where winners receive the right to purchase tickets. This eliminates the need to line up or spend hours on the telephone trying to buy tickets when they go on sale. The lottery winners then have a certain amount of time to purchase the tickets (typically one week) before their rights are canceled.

Although all three-ticket sellers provide this lottery service to users, Pia and Lawson have created Internet and telephone purchasing processes that make the best of their ticket outlets. Users can receive a code number when they reserve a ticket on the Internet or on the phone. In the case of Lawson, users input this code number into a kiosk in a Lawson convenience store to receive a ticket, which they pay for at the register. Pia uses a similar system in its ticket outlets and in conjunction with convenience stores.

The use of these code numbers and the delay between reserving and paying for tickets can cause problems for ticket sellers and promoters since some people, particularly scalpers, will register for many lotteries under various names and thus win the rights to purchase many tickets. If the concert is popular, they sell the tickets for very high prices, which can annoy the real customers who merely want to see the concert. The bigger problem occurs when the concert turns out not to be popular and the scalpers do not exercise the right to buy tickets, thus leaving many tickets unsold at the last minute.

Electronic Tickets

Concert promoters believe that smart cards or phones that contain infrared technologies or smart card capabilities will solve or at least partially solve the scalper problem since they can assign tickets to individuals as opposed to code numbers when the tickets are sold. In the case of smart cards, the ticket sellers will distribute the smart cards to their registered members and assign the tickets to these smart cards when a member reserves a ticket. Like the train tickets, the ticket checking machines compare reservation and member information when the member passes the smart card over the appropriate place on the ticket machine.

Electronic tickets can be beneficial to both ticket seller and promoters. They reduce transaction and delivery costs for ticket sellers and enable promoters to increase the amount of same-day sales. Since promoters do not want to be left with unsold tickets, they will probably be willing to discount tickets at the last minute. Instead of having radio stations give away unsold tickets, the Internet (in particular the mobile Internet) and smart cards will make it possible to vary the price of the ticket according to the demand conditions as the concert date approaches. The mobile Internet makes this even more of a possibility since buyers can presumably respond much faster to announced-price reductions in mobile mail services than PC users.

Electronic tickets also enable firms to further replace direct mail marketing with Internet mail marketing. Just as most products purchased over the mobile Internet are chosen through mail as opposed to search services, most concert tickets are purchased in artist-specific mail services. Concert-ticket providers can use these mail services in combination with the smart cards to send surveys or advertisements for music paraphernalia to concertgoers right after the concert is over. Internet reservations enable some of these mail services, but smart cards further connect the ticket to the individual and thus promote greater Internet marketing.

Japanese concert ticket providers have been experimenting with 2D bar codes, infrared-compatible phones, and smart cards since 2001 and had not made a decision as of mid-2003. In addition to many of the issues mentioned above, they are also grappling with how to assign multiple tickets to multiple people when a reservation is made for multiple people. Since few people attend a concert by themselves, it must be possible for one person to make reservations for multiple people and then have the tickets assigned to multiple people. Imaginative scalpers may be able to use this weakness or other weaknesses to continue their business.

A second stumbling block is how to divide up the benefits from electronic tickets and the costs of implementing them. The greater use of Internet reservations will probably mean increased credit card fees along with costs for implementing readers, writers, and, if applicable smart cards. Concert promoters would most likely bear many of these costs; they would have to install readers and writers in each concert location or buy portable ones that could be moved to the different concert halls. On the other hand, ticket sellers might bear the cost of the physical smart cards.

Other Entertainment Tickets

It is also possible to use smart cards or phones that contain infrared technologies or smart card capabilities in other entertainment events such as sporting events and movies. As with the concert tickets, Internet reservations in combination with electronic tickets can reduce transaction and delivery costs and enable more same-day sales of other entertainment tickets. Similarly, the various participants in the system must agree on how to divide up the benefits and costs of implementing smart cards.

Movie tickets present a unique set of problems. As with concert tickets, the United States is ahead of the rest of the world in online sales of these tickets where more than 3% of movie tickets were bought from online from sites such as Moviefone, MovieTickets.com, and Fandango as of late 2002. After making reservations on a site, users either print out the tickets or pick them at the movie theaters right before the movie starts. The former requires movie theaters to read the tickets, probably in the form of a bar code, with scanners[3].

Allowing users to make these reservations on phones provides users with an additional amount of convenience. Not only will mobile services probably attract those people who prefer the mobile to PC Internet, it also accommodates those PC Internet users who do not plan their entire day before they leave the house. These people can make their reservations while outside their homes and then use their smart card or phone as a ticket. Due to the low price of movie tickets, it might be possible to send bar codes to phones and then scan them with readers like Jeansmate does with discount coupons.

In the end, the popularity of these services will probably depend on how crowded movie theaters are. If people can get seats without making reservations, they will probably not bother with reservations. Some people have argued that online movie ticket sales will not succeed as long as there is an oversupply of movie theaters, a condition that can quickly change[4].

An alternative is to offer reserved seats and charge an additional fee for these seats. This would enable some people to arrive just before movie starts and still get a good seat. Other entertainment events do it like sporting events and concerts, so it is natural that movie theaters do it. Many Japanese movie theaters have started to do this with the expectation that Internet reservations will increase. The biggest challenge might be history and the expectation of consumers, who expect to be able to sit anywhere.

Additional fees would naturally improve the profitability of the online sites in the United States, which are currently struggling to make a profit. Apparently some sites make most of the money from advertisements as opposed to transaction fees[5]. This is understandable given the low prices and commissions with movie tickets and the high fees for credit cards.

MONEY

Currently, cash is primarily used in small transactions and here is where smart cards, particularly non-contact smart cards, have an advantage over credit cards, debit cards, and checks. Non-contact cards are much faster, and they have lower transaction costs (almost zero) than all of these alternatives, including debit cards.

The main challenge for smart cards with money is solving the network effects or chicken and egg problem. Although similar problems exist with transportation and concert tickets, in those cases there are large firms who can single-handedly distribute the cards and install the readers. National railways, like JR in Japan, can distribute cards to its riders and put the readers in its train stations. Large concert ticket providers can distribute cards; the concert promoters, who are their partners, can install the readers.

In the case of money, there are no firms who can both distribute cards and implement the readers. Some expected the credit card companies to play the former role while restaurant and hotel chains played the latter role. Credit card companies have been pushing contact cards for many years. But restaurants and hotels aren't very interested in purchasing new readers particularly if the credit card companies will not offer them lower fees in return for implementing the readers. Recently credit card companies have also started pushing infrared technologies in Japan and Korea where the infrared device in the POS register enables an easy connection to the existing credit card network.

Convenience Stores

Convenience stores are moving the fastest to introduce smart cards in Asia; Hong Kong appears to be the leader, with Japan struggling to catch up. The convenience stores have national networks to help them solve the network problem, and the characteristics of smart cards better match their needs than the needs of other stores. The low purchase amounts reduce the problems of stolen cards and thus makes security in the forms of signatures and other identification unnecessary. They have the high volumes needed to justify large investments in readers, and most importantly they are very concerned about processing times.

In fact, Asia's convenience stores may have highest sales per square foot than any other type of store in Asia, including Japan, and in the world. They are located in the areas of Japan that have the highest amount of pedestrian and car traffic, and they make money purely through volume. The most successful convenience stores are near train stations, where the pedestrian traffic is the highest. Their goal is to process payments as fast as possible in order to keep the lines short and encourage more people to come into their stores.

Many of Japan's conveniences stores also want to introduce these smart cards in order for its customers to automatically pay for other in-store services. This includes in-store printing services and special-purpose terminals that they originally hoped people would use to do Internet shopping. Although the latter application has not taken off, Japan's convenience stores have integrated their mobile Internet services with these terminals (e.g., the above discussion of concert tickets), and they would now like to integrate smart cards with these terminals.

Japan's convenience stores also have allies in the form of Japan's service providers. With more than 70 million subscribers, the service providers have the power to distribute these smart cards as a function in its phones to large numbers of people. They can also advertise this capability on prime time television and other media. Many convenience stores expect payments to begin shifting from cash to phones as the service providers do these things. For the convenience stores, one question is, what percentage are they willing to pay the service providers, or a smart card company, in order to speed up payments at the register and promote other in-store services?

Higher Denomination Purchases

One reason why Japan's convenience stores and Japan's service providers are optimistic about these non-contact smart cards is the success of debit cards in many U.S. supermarkets. U.S supermarkets began accepting debit cards in the late 1980s because they wanted to eliminate checks, which have high transaction costs particularly when they are returned, and because debit cards have lower transaction costs than credit cards. The existence of credit-card readers and the credit-card payment system (both are used for debit cards) in supermarkets facilitated the introduction of debit cards, and the diffusion of ATMs in the early 1980s caused the diffusion of PIN numbers. Simply by introducing PIN pads, supermarkets were able to easily accommodate debit cards.

Japanese firms never introduced debit cards, partly because credit cards have diffused slower in Japan than in the United States and also because Japanese banks are not known for their innovativeness. However, the success of debit cards in the United States suggests that convenience store and supermarket customers may be willing to use smart cards in Japan and perhaps even in the United States.

Promoters of non-contact smart cards in Japan hope they can expand the applications from transportation and entertainment tickets to high denomination purchases and thus replace debit and credit cards since they have lower transaction costs and they are much faster than both debit and credit cards. For example, Japan's largest railway company, JR, which is the largest user of non-contact smart cards in Japan, has combined its credit card and smart card into a single card. It hopes to use the popularity of the transportation cards to gather more credit card customers.

These kinds of activities threaten large credit card companies like Visa and MasterCard who have created a network of merchants and banks. This is one reason why Japanese and Korean credit card companies are trying to have their cards used in transportation and other applications in addition to their efforts to promote infrared techniques.

They recognize that the emergence of a widely used non-contact smart card, particularly one that was inside of a phone, would present an alternative to existing credit cards. By merely requiring signatures and identification, non-contact smart cards could provide similar levels of security to existing credit cards (either magnetic or contact smart cards). To understand how the success of non-contact smart cards might threaten the credit and debit card industry, we need to take brief look at this industry.

Credit and Debit Cards

A credit and debit card payment network involves merchants, cardholders, credit card companies, and, in the case of open-loop systems, acquirers and issues (see Figure 8.1). American Express and Discover have closed-loop systems because they have contracts with all cardholders and merchants that belong to its system, and thus all transaction data are captured within the system. Visa and Mastercard utilize acquirers to sign-up merchants and issuers to sign-up consumers, and thus data must be transmitted among thousands of firms[6].

The existence of acquirers and issuers makes it easier to create open-loop systems than to create closed-loop systems, and closed-loop systems are more flexible than open-loop systems because individual issuers can develop and offer individual features for their cards. On the other hand, the open-loop systems are more complex in terms of the authorization and settlement system and thus involve higher transaction costs. Closed-loop transactions are typically just routed between the merchant and the card system, which has all the information on the cardholder.

These payment card networks have substantial fixed costs. This includes the infrastructure to process the transactions, national advertising, and a merchant and cardholder base. For the latter, in the mid-1990s it cost banks about $53 to sign up a cardholder. These high fixed costs are a major reason why most of the early entrants lost substantial amounts of money and exited quickly. Even the successful firms experienced difficult times in the industry's early years.

Credit card companies and their acquirers and issues primarily make money through merchant discounts and interest charges on unpaid bills. The former has steadily decreased from 7% in the 1960s to 1.8% as of the late 1990s. In the open-loop systems, the acquirers make about 0.4% and the credit companies share 1.4% with the issuers. Even if the transaction costs for the issuers is higher than for the acquirers, it is

FIGURE 8.1. Credit Card Payment System Network

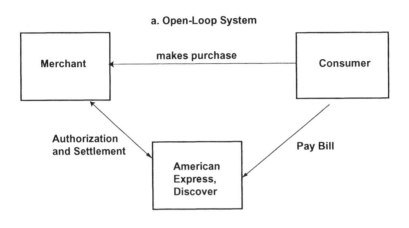

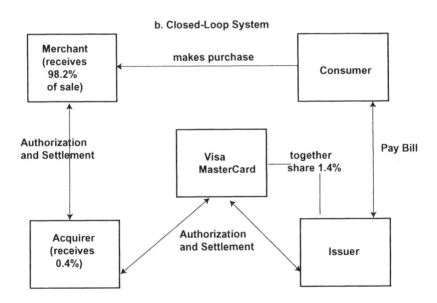

Source: Adapted from Evans, D. and R. Schalensee, *Paying with Plastic: The Digital Revolution in Buying and Borrowing*, Cambridge, MA: MIT Press, 1999.

clear that the credit card companies are making quite reasonable profits.

The lower transaction fees that credit card companies charge supermarkets also suggests that the profits for credit card companies are large. Credit card companies generally charge stores about 0.4% less for credit card transactions than they do for restaurants in spite of the fact that their costs are relatively unchanged. For debit cards they introduced a flat rate of about $0.08 per transaction, partly because debit transactions have lower costs than credit card transactions (the money is deducted from the user's account as opposed to providing the user with a short-term loan). Another possible reason that credit card companies introduced these lower fees for supermarkets was to attract a new set of customers who are merely more price sensitive than their main customers[7].

These high profits are one of the reasons why new entrants such as non-contact smart card providers and the mobile service providers want to have their cards and phones used as credit cards. The use of smart cards in transportation, entertainment, and convenience stores will probably reduce the cost of smart card readers. The service providers merely have to set a lower transaction fee than the credit card companies have done to attract stores to their networks, and they hope that the convenience of the phone will attract users. With transaction costs for smart cards near zero, the service providers can drastically reduce the transaction fees. This might force the credit-card companies to release their transaction fees, which by itself would be a major blow to them.

Furthermore, the service providers may have a number of other advantages. First, they can sign up cardholders much more cheaply than the banks did, given their strong brand images in most countries. Second, by putting the credit card inside the phone, consumers may no longer know whose credit card they are using. They may know the phone manufacturer and the service provider, but they may no longer know the name of the credit card company.

BIOMETRICS

Another way that smart cards or phones containing these smart card functions might replace credit and debit cards without requiring time consuming signatures and identification checks is through biometrics. Biometrics refers to a group of technologies that verify individuals in accordance with their measurable physiological traits. For reasons completely independent of the mobile Internet, many firms are developing technologies for automatically identifying fingerprints, palm prints, hand geometry, vein patterns, retinas, and other physical traits. By placing the appropriate template in the phone or server and the appropriate reader in the phone, it becomes possible to check and authorize the user with a fairly high level of reliability, thus eliminating time consuming signatures and identification checks during credit card purchases[8].

For example, NTT DoCoMo introduced a phone with a fingerprint reader in the summer of 2003. Other firms have developed technology for authorizing users based on facial (Omron) or voice (Matsushita and KDDI) characteristics. One key issue in these systems is whether the template is contained in the phone or the server. While current systems like the fingerprint reader in NTT DoCoMo's phone have placed the template in the phone, placing the template in the server would add additional security but increase the processing time. A second issue is the amount of processing power needed for the fingerprint or other reader and in the case of storing the template in the phone, the amount of memory for storing the template. As the processing power and memory in phones increase, it will become possible to reduce user authorization times thus increasing the chances that phones in combination with smart card functions will replace credit cards.

Biometrics might also make it possible to use smart cards or phones containing these smart cards for more expensive tickets like airline tickets. Most airlines have been pushing electronic tickets since 2002 where people print out the tickets from their PC and then input ticket numbers and/or credit cards in a kiosk at the airport. By placing the appropriate readers at airport check-in counters, it would be possible to use smart cards or phones with these functions as the airline tickets. Of course, they could also be used for security checks, a fact that governments have already realized. The use of phones as inexpensive tickets and money will increase government interest in using them as national identification cards.

OTHER APPLICATIONS

Many people in Japan argue that the best way to forecast the future applications for phones is to empty your pockets on the table. We have already covered tickets and money in this chapter and covered loyalty and business cards in chapter 5. National identification cards, medical insurance cards, and keys are additional examples. We can already unlock our doors with an infrared device that is attached to our keys. It would be simple to place this infrared device in our phones.

It might also be possible to apply the same concept to homes and buildings. Already many firms use smart cards for preventing access to buildings by unauthorized people. Perhaps it is possible to apply the same concept to hotel rooms and homes. Room keys could be given to phones through the use of the smart card function in the phone or more simply by using the infrared functions when people check-in to a hotel.

Changing the locks on your homes or apartments can be expensive, but it is likely that at some point the trend will start. It could be accelerated by the recognition that many buildings have very insecure locks. The locks for many buildings have been designed such that anyone can create a master key for an entire building by using any key from that building[9]. If we are to redesign locks, why not do it with phones?

SUMMARY: THE END OF CASH?

It is possible that phones will eliminate physical cash within the next 10 to 20 years and become one of the major sources of payments in the world. While the initial changes may be slow, as the cost of readers decline and consumers become accustomed to using their phones for tickets and small payment, the pace may likely accelerate as a critical mass of users is created for each application. Increased processing speeds will also increase the use of biometrics, which will facilitate the use of phones for higher denomination purchases. Many firms will be affected by these changes, credit card companies and banks are a few examples. Who needs branches and ATMs when physical money no longer exists?

Some people believe that it is in government's interests to eliminate cash. These proponents estimate that the handling of cash represents more than 1-2% of America's GNP. More importantly, cash is the basis of most drug dealing and terrorist activitiesl; the former may represent more than 10% of America's GNP. While the elimination of cash would not eliminate these activities, it would make some of these activities more difficult and increase the importance of government efforts to eliminate bank secrecy[10]

The elimination of cash would also make it easier for governments and other institutions to monitor our actions. If most monetary transactions were captured electronically, it would become easier for governments to monitor people's income and collect taxes. It might also improve the reporting of firm performance as more transactions are captured electronically.

Discussing the end of cash and even the use of smart cards for something as simple as transportation tickets brings up the issue of privacy. Most of us want the government to catch the big tax evaders but not our small errors of omission. But with the spread of smart cards and the Internet, governments should consider the possibility of eliminating cash and the ways that it can be done in a way that minimizes its effect on our privacy and individual rights.

NOTES

[1] Evans, D. and R. Schmalensee, *Paying with plastic: the digital revolution in buying and borrowing*, Cambridge, MA: MIT Press, 1999. Much of the information about the United States market that is presented in this section is drawn from this book.

[2] The phones do require a larger signal (plus battery power) due to the interference among parts inside the phone and the existence of the shield, problems that don't now exist when someone places their smart card over a reader.

[3] Miles, S., "'Star Wars' Sequel Could Mean Record Sales," Headaches Online, *The Wall Street Journal*, May 10, 2002.

[4] Miles, S. "'Star Wars' Sequel Could Mean Record Sales," Headaches Online, *The Wall Street Journal*, May 10, 2002.

[5] King, T., "Hollywood Journal: The Battle for Your $1 Advance Movie - Ticket Sales Heat Up This Season; Call Ahead for Popcorn?", *The Wall Street Journal*, December 21, 2001.

[6] Most of the information in this section is taken from Evans, D. and R. Schmalensee, *Paying with Plastic: the digital revolution in buying and borrowing*, Cambridge, MA: MIT Press, 1999.

[7] Academics typically use the term "personalized pricing" to describe this phenomenon, which is the same concept used to justify discount coupons.

[8] Wilson, C., *Get Smart: The Emergence of Smart Cards in the United States and their Pivotal Role in Internet Commerce*, Richardson, TX: Mullaney Publishing, 2001.

[9] Schwartz, J., "Master Key Copying Revealed," *New York Times*, January 23, 2003.

[10] Warwick, D. *Ending Cash*, NY: Quorum, 1999 and Schlosser, E., *Reefer Madness: Sex, Drugs, and Cheap Labor in the American Black Market*, NY: Houghton Mifflin, 2003

Chapter 9

Mobile Intranet Applications

Mobile Intranet services may represent the largest potential market for Internet-compatible phones. As firms build and expand their corporate Intranets, the need for access from mobile phones will increase. Many Japanese business people already perceive that it is more convenient to maintain schedules and handle mail on their mobile Internet-compatible mobile phones than on PDAs. And as the performance of mobile phones increases, mobile phones will be used for more sophisticated applications.

The large market for mobile phones drives this virtuous cycle. More firms create new technologies for the mobile phone than for the PDA because the market for mobile phones is more than 30 times larger than the market for PDAs. The larger number of new technologies will cause mobile phones to be improved at a faster rate than PDAs. The subsidies that most service providers in the world provide for phones reduce the cost of phones to users, and thus also drive a faster improvement rate for phones than PDAs. And most importantly, since phones are starting from an inferior position, the same percentage improvement in phones and PDAs will have more meaning for phone users than PDA users. As argued in Chapter 1, PDAs may eventually play the same role in the mobile environment as workstations do in the desktop environment, always better than mobile phones but never the main form of access like mobile phones.

The concept of disruptive technologies will play an important role in the diffusion and improvement of these mobile phone-based mobile

Intranet systems. The mobile phone's small screens and keyboards make it hard for companies to use mobile phones in the same way they have used PCs in their Intranets. For example, it is likely that the PC will continue to be the main form of access for B2B applications (other than some of the restaurant and bar examples presented in Chapter 7) for the near future. Furthermore, similar reasons may also cause ERP (Enterprise Resource planning), CRM (Customer Relationship Management), and to a lesser extent SFA (Sales Force Automation) to play a small role in the early years of mobile phone-based mobile Intranet systems. Many suppliers of these ERP and CRM systems are trying to help their current customers in Japan access company Intranets from outside the office with both mobile phones and PDAs through improved software packages and as an Application Service Provider (ASP). Nevertheless, it appears that most of their customers are more interested in perfecting access from their PCs (and to a lesser extent from PDAs) than in making the systems accessible from mobile phones.

Instead it is a different set of users, applications, and suppliers who are driving the implementation of mobile Intranets in Japan. Firms that have little experience with ERP and CRM are implementing a new set of applications and they are using a new set of solutions that are offered by a new set of firms. More than 50 firms were offering almost 80 mobile business solutions in mid-2003, and none of these solutions were from the leading suppliers of ERP and CRM like SAP, PeopleSoft, Oracle, J.D. Edwards & Co., or Siebel.

Furthermore, most users are doing their own systems integration and are probably providing a larger percentage of innovations than the suppliers of the software systems do. Just as users were the main sources of innovations in the first computer systems[1], scientific and measuring equipment, and various kinds of manufacturing equipment[2], many firms are developing their own mobile Intranet systems because they have needs that are not yet understood by suppliers. The mobile nature of the devices places unique demands on the system, the small screens and keyboards place unique restrictions on the type of information that can be accessed, and improved applications processors are rapidly changing the phone's user interface. Until suppliers understand these unique demands and restrictions and the true capabilities of phones become clear, it will be difficult for them to offer standard systems that can be used by a wide variety of firms.

Table 9.1 summarizes three stages of mobile Intranet usage; mail, the first stage, can also be divided into three stages. The first section of this chapter describes some simple mail solutions and examples of these systems in applications like delivery, taxis, and trucking. The second

TABLE 9.1. Growth in Corporate Users for Three Stages of Mobile Intranet Usage

Stage	Applications	Number of firms		
		2000	2001	2002
1 Mail				
1.1 Forward	Forward PC mail to phone	6	25	33
1.2 Instructions	Send instructions to phones	5	31	52
1.3 Secure mail server	Forward mail or instructions via secure mail server	6	17	51
2. Groupware	Enable groupware access on mobile phone	22	43	87
3. Access company databases	Access and input sales, product, price, customer, and order data	14	37	101

Source: Japanese firms and author's analysis

stage is mobile groupware and the third stage is accessing corporate data from the mobile phone. The latter sections of this chapter discuss mobile groupware and some examples of Stage 3 systems in applications such as maintenance, construction, and sales force automation (SFA). Some of these Stage 3 systems rely on mail and thus are called push-based systems where the addition of URLs to these mail messages makes them Stage 3 systems.

Using data from various Japanese firms, we can also estimate the number of firms and business people who are using phones to access corporate data. Table 9.1 shows estimates for the number of those firms that had corporate accounts with service providers, which had reached the three stages as of the end of 2000, 2001 and 2002. As shown in Table 9.1, growth occurred primarily in Stage 1 systems in 2000 and 2001, while growth centered on Stage 2 and 3 systems in 2002. It is likely that growth in Stage 3 systems will accelerate in 2003 and continue to exceed that of Stage 1 and 2 systems.

The same data suggests there are at least 10 firms that have introduced Stage 3 systems in which there are at least 3000 busness users in a single system. This suggests there are probably more than 100,000 users of these systems and the number is growing very fast. It is possible that this number will exceed one million by the end of 2005.

Accelerated growth in these Stage 3 systems will likely drive further increases in corporate data-usage[3] and possibly cause business usage to exceed entertainment usage within 3-5 years.

STAGE 1 SYSTEMS: MAIL

There are a variety of ways to handle corporate mail. The most simple solution is for individuals to have their PC mail forwarded to their phone. Potential problems with this approach include viruses, crowded in-boxes on the phone, and full mailboxes in the i-mode servers. Unless they are almost never in their office, most people do not want to have all their PC mail messages arriving on their phone each day. Scrolling through tens or even hundreds of mail messages can be very time-consuming.

Furthermore, server problems caused many Japanese firms to temporarily forbid this approach. Service providers set limits on the number of mail messages that are saved in their servers, and full mailboxes typically generate a mail message asking the user to delete messages. There were many cases where these messages started an infinite loop that resulted in severe system problems for firms.

Net Village

Net Village provides a simple solution to this problem called remote mail service. For several dollars a month, individuals can have their PC mail converted to c-HTML format so that it can be read on their mobile phone. Users first register their PC mail server and user names along with the password. When they wish to access their PC mail from their mobile phone, they access Net Village's mobile site, input the password via the mobile phone keyboard, and the mail is loaded onto temporary home pages. The phone simultaneously displays the titles of five messages; placing the cursor on a title causes the full title to scroll across the page, thus revealing the full title. Clicking on the title downloads the full mail message. Functions include address book management and the ability to refuse specific mail addresses or only have certain mail addresses downloaded.

Net Village's Java program provides additional capabilities. By eliminating the need for downloading the tags and other formatting information, the Java program reduces the packet charges by more than 50% while at the same time providing faster mail response. Net Village's servers will check the PC mail in-box as often as every 30 seconds for mail including the reporting of mail from specific mail addresses. Furthermore, it enables users to activate other Java programs such as screen savers (which could be corporate-based) and maps and even access Microsoft Word and Excel files.

A key issue is the expandability of such a service that is currently aimed at individuals. Many organizations (including my university) do not allow outside access to their servers, and for this and other reasons Net Village's may always be a solution to an individual's mail problem. Increasing the size of the Java program will increase the complexity of applications, but will multiple people be able to share information? Although there are currently about 350,000 subscribers, this figure passed the 300,000 mark in late 2001 and subsequently has seen very little growth. Therefore, the percentage of total mobile Internet subscribers in Japan who are subscribing to Net Village's service has actually declined over the last two years. Is this because the service is not providing enough value to its users? Or is it because firms do not perceive a future for these services?

Secure Gateways

Security is a key issue in mobile Intranets, and firms are developing a variety of ways to handle it. The most simple method relies on users inputting user names and passwords. Of course, user name and passwords can be stolen, and one option is to constantly change these user names and passwords.

More than 10 firms offer more sophisticated approaches. For example, IDS (International Digital Solutions) uses a second firewall in addition to the one that relies on user names and passwords. The additional firewall is the sending of mail, which includes a URL, to the authorized mail address following the authorization of the user's name and password. Thus, even if the user name and password are stolen, mail containing the appropriate URL is only sent to the authorized mail address. Furthermore, since a random number generator changes the URLs each time they are sent to an authorized user, the URLs by themselves are useless. Therefore, unless the user name, address, and phone are all stolen, unauthorized people cannot enter the firm's mobile Intranet.

IDS (International Digital Solutions) has received a patent for its system in the United States and Europe and is waiting for a patent in Japan that it applied for in mid-2001. More than 20,000 business people were using systems based on IDS's technology as of June 2003. This includes Sony's SFA solution, which is discussed at the end of this chapter. A number of other users rely on similar concepts in their systems. For example, one of the reasons why firms are using push-based (examples are discussed in the following sections) as opposed to pull-based systems is that push-based systems only send mail to authorized users.

A futuristic method is to use biometrics, which was discussed in the previous chapter in terms of tickets and money. As phones include the technology for comparing the user's fingerprint, facial, or voice characteristics with a template, firms can use such technology to confirm a user's fingerprint, face, or voice before allowing the user to enter the system. While current systems like the fingerprint reader have placed the template in the phone, placing the template in the server would add additional security while perhaps increasing the processing time.

EXAMPLES OF STAGE 1 SYSTEMS

Stage 1 systems rely on mail, which could be personal mail forwarded from a PC or orders/instructions sent to employees. This section describes examples of the latter systems. Although SMS messages can also be used to send such orders and instructions to employees, it is easier to add URLS to Internet mail than SMS messages.These URLs enable employees to access and input more complex information, thus turning a Stage 1 into a Stage 3 system. It would be very easy for the firms discussed in this subsection to change their Stage 1 into Stage 3 systems.

Delivery Applications

Delivery companies are some of the leading users of information technology in the world. Firms like UPS and Federal Express have been using GPS and sophisticated information systems to manage the pick-up and delivery of parcels for many years. For example, UPS introduced special-purpose wireless handsets as long ago as 1993 that enable it to transmit delivery information and customer signatures from the company's 50,000 vehicles to the UPS mainframe. Previously, this information was not available until the day after delivery. UPS also allows its customers to access this information through the Internet[4].

The mobile Internet and conventional GPS systems provide new options and new tradeoffs for these delivery firms. For example, Sagawa Kyubin began using mobile phones instead of proprietary handsets (a customized PDA) for communication between call centers and drivers in early 2001. Operators contact drivers with mobile mail concerning changes in pick-up or delivery locations and times, and the drivers simply click on a URL and the appropriate place within the home page to signal that the delivery has been completed. The central database is updated every 15 minutes. The implementation of the system, including the 25,000 phones, cost 2 billion yen (about $150 million).

Bike Dot Japan began using conventional GPS and mobile mail in place of their proprietary mail systems (a customized PDA) to communicate with delivery personal in late 2001. Bike Dot Japan delivers parcels using small motorcycles. Operators in the call center look at computer screens that show the pick-up and delivery points for the parcels to choose the most appropriate vehicle and notify it with mobile mail. It is now possible for one person to complete the job whereas previously it took one person for the order and one person for the delivery notification. Using this technology Bike Dot Japan has been able to reduce call center personal needs by 40% and delivery times by 5 - 10 minutes.

A third Japanese delivery company, Yamato Express, still uses its proprietary handsets to exchange delivery information with the host computer. Yamato delivers about 2.5 million packages per day using a kind of hub-and-spoke system. The system includes 2300 hubs and 310,000 spokes; most of the spokes are convenience stores from where customers send packages. Yamato's drivers use a proprietary handset that includes a wireless MCA card and a bar code scanner to transmit delivery information to the host computer.

Unlike the other two examples, Yamato Express uses the mobile mail function to improve customer service. Like UPS, users can access information about their package on their PCs and mobile devices; Yamato Express has added the mobile phone to this list of devices. In addition, users can set the delivery time from both their PC and mobile phone, the latter of which is much more important for people who want to change the delivery time. If they had access to their PCs, they would probably be at home and thus could take the delivery! After choosing a delivery time, users receive a confirmation mail from Yamato. Changes must be made by 6 A.M. of the delivery day and deliveries can be delayed up to one week.

Yamato will also inform customers via a mail message in the case of late packages. If the sender registers the receiver's mail address, Yamato can also inform the receiver; this has been available since early 2002.

Yamato hopes that these kinds of mail services will increase the percent of cases in which someone is home when the delivery arrives. Repeated attempts to deliver a package can quickly increase a delivery company's costs. Reducing the numbers of multiple deliveries will depend on both new technologies and on changing customer behavior; the second task may be the hardest.

Taxi Companies

Taxi companies are another potential market for mail-based systems. Historically, call centers receive calls from customers and use radio systems to find the closest free taxi; the latter can be very labor-intensive and often inaccurate. Conventional GPS systems, like the ones used by UPS and Federal Express, and phone-based location (based on either the closet base station or GPS) systems can provide operators with information about taxi locations. Mobile mail can provide a new form of communication between call centers and taxis and make a "virtual" taxi company possible (see Figure 9.1).

A number of Japanese and other taxi companies have begun using conventional GPS or phone-based systems and mobile mail to improve operations. For example, a Kyoto-based taxi company uses a conventional GPS system to display taxi locations on a PC screen. Operators use their wireless radio systems to contact the closest taxis to find a free one. They also reduce competition between their taxis by preventing multiple taxis from serving the same area. A second step is to have taxi drivers update their status with mobile mail, thus enabling the system to choose the taxi driver with lower manual input and send them mail concerning the destination.

Taxi cab drivers may also begin using the mobile Internet to search for places where demand for taxis is high. Demand for taxis is often high following a concert, baseball game, or other event. While taxi cab divers have historically searched for this information on their own in newspapers and magazines, ticket sellers like Pia Ticket have begun offering this information to taxi cab drivers via a mail service. GPS-compatible phones can receive information on events that are near their location. In the long run taxi cab companies may want to use this information to dispatch specific taxis to the event.

On the other end of the system, some taxi companies have been taking requests via both the PC and mobile Internet for many years, which is the third step shown in Figure 9.1b. Currently, many users of these services have speaking, hearing, or other problems that make it easier

FIGURE 9.1. Taxi Applications

a. Virtual versus Existing System

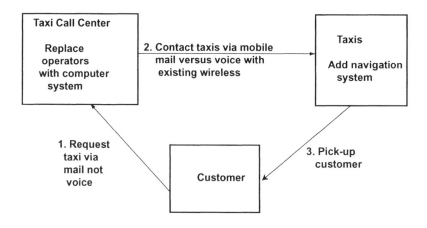

b. Six Steps Towards the Virtual Taxi Company

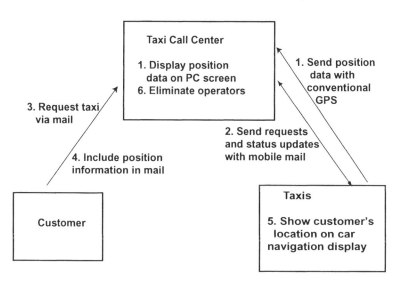

to reserve taxis with Internet mail than with a phone call. Users input their name, location, and phone number on a PC or mobile Internet home page. Operators then search for the closest free taxi on their PC screen or in via voice on their wireless system and then send the user mail containing the taxi driver's phone number. The customer and the driver can set the appropriate meeting place and time.

A fourth step is to estimate the caller's location using either the closest base station or the GPS function in phones. A number of Kyoto-based taxi companies began offering such services in 2003 including emergency services. The companies are targeting physically impaired and elderly people with these services. Fifth, Japanese and European taxi companies are replacing their conventional GPS systems with more sophisticated car navigation systems so that taxi companies can display the customer's location on the taxicab's car navigation display[5].

These technological trends and the high costs associated with them may also cause the emergence of virtual taxi companies that do not own taxis but merely handle the requests. Several Japanese companies have started offering ASP services that provide some of the services just mentioned. As these companies improve their services, the possibility of a virtual taxi company increases. Customers request a taxi through PC and mobile Internet connections; the request, including the location of the customer, is automatically sent to the closet free taxi via mobile mail. Taxi drivers confirm their intention to pick up the customer and the actual picking up and delivering of the customer via a URL in the original mail message. The virtual taxi company uses this information to update the status of each taxi.

It is also quite likely that some aspects of such virtual taxi systems can be used to create virtual trucking systems. Trucking systems also require truck drivers to receive instructions about their next pickups and thus mobile mail and GPS can be used in a manner similar to that of the taxi systems. Differences between the trucking and taxi applications include the need for unloading and loading physical products and documenting these activities. Chapter 7 described one-way mobile phones in combination with conventional GPS systems that can be used to handle the documentation of these activities.

STAGE 2 SYSTEMS: GROUPWARE

Mobile groupware can be defined as the second stage in the creation of mobile Intranets since it can provide more functions than the mail-based services described above. Groupware started with software packages such as Lotus Notes and Notes Exchange and now a large variety of these packages exist, many of which address specific applications such as project management for a specific industry. In addition to mail, users can access project schedules, bulletin boards, and other people's calendars and share files. And many of these systems can be downloaded and accessed over the Internet. In Japan, new firms such as Cybozu and Dream Arts were the first firms to introduce the Internet-compatible packages, but the large incumbents such as IBM and Microsoft have subsequently introduced similar systems.

Mobile packages

More than 20 firms offer mail and groupware solutions for mobile phones. Some of these solutions provide compatibility with existing PC software such as Microsoft Outlook. Other solutions focus on specific tasks such as instant messaging or exchanging multimedia image or more general applications such as sales force automation, time card or bulletin board services, university clubs, or community sites.

Two of the leaders are Cyboz and Dream Arts. Cybozu introduced software in January 2001 that includes the ability to access some groupware functions such as mail, calendars, and bulletin boards on mobile phones and PDAs. The software recognizes whether the access terminal is a phone or a PDA and adjusts the screen sizes accordingly. The PDA version also has browsing and data synchronization functions and accesses more data in a single download and saves it on the PDA so that it is faster for the user to retrieve the data. Although the market for this mobile groupware initially soared at one point, reaching 20% of total sales for Cybozu, it subsequently fell back to several percent of Cybozu's sales in 2002. Apparently most people find the calendar and bulletin board functions difficult to use on the mobile phone's small screens, and mail is the only function that is regularly used.

Dream Arts also offers a mobile groupware package. While Cybozu targets small groups that contain as few as 10 people, Dream Arts targets much larger groups; up to 10,000 people can use its groupware program. It sells these packages through system integrators such as IBM while Cybozu's software can be downloaded from the Internet. Dream Arts' focus on larger groups has caused it to focus on applications that are slightly different from those of Cybozu. One function enables users to create and send surveys to mobile phones. While it is also possible to send surveys to PCs in traditional groupware packages, there are critical differences between PC and mobile phone surveys as discussed in Chapter 5. Furthermore, users can also input data into their mobile phone while interviewing consumers where the data could be text, voice, or pictures. Camera phone users can attach pictures to mail and send them to their PC where the Dream Arts' system enables users to create reports and schedules from these inputs.

Like the type of services offered by Net Village, there are a number of questions about the future viability of mobile groupware packages. Although their focus on group as opposed to individual problems suggests that the market for groupware will quickly surpass the market for Net Village's services, how can mobile groupware suppliers use new technologies such as Java to expand the functionality of their products? Can groupware be constructed from a set of smaller client-based programs such as Java or will existing suppliers of PC-based groupware resist such efforts for either technical or organizational reasons? Without such a capability, groupware cannot benefit from the increased processing power of phones and thus could be at a disadvantage when compared to new approaches.

A few firms have started to introduce Java-based groupware solutions; one of these firms is NTT Software. NTT Software has introduced an application service called Go Wireless that enables users to access groupware packages such as Microsoft Exchange via a Java program and the application service that is housed in a server. The Java program reduces the amount of tags and other formatting information and thus reduces data charges. Furthermore, it includes a pop-up menu thus improving the user interface. Although the version released in late 2002 only provides users with access to groupware functions such as mail access, address management, and schedule management, future versions of Go Wireless are expected to provide access to CRM and SFA programs.

Multi-tasking

There are also fundamental changes in the phone design that will facilitate solutions such as mobile groupware. On the PC, many people simultaneously work on Microsoft Word and Power Point documents while they keep their Microsoft Exchange or Eudora mail program running. They refer to multiple documents while they work and exchange data between the documents.

Existing mobile phones cannot do this due to restrictions in the operating systems that are used in mobile phones. For example, in the maintenance examples described below, the maintenance worker cannot easily move back and forth between viewing a mail message or home page, or making a call. The worker must cancel the mobile Internet session in order to view the mail message again or make a call (this problem does not exist when a phone call comes in[6]). This makes it difficult to refer to the information in the mail message or in the URL while talking to someone.

Furthermore, it is even difficult to return to a home page after making a call. In the maintenance example, it is possible to bookmark the URL before making a call, although this requires about the same number of clicks as re-accessing the URL through the mail message. With official sites, URLs are not available and thus the page cannot be saved. Therefore, returning to the site requires the user to pass through the main menu and sub-menus again, clearly a time-consuming set of activities.

A simple solution is to use a Java program to manage various applications including the downloading of data into these applications. For example, a firm called Gravana offers Bullant Remote, which is a 20-kilobyte Java program for managing multiple applications simultaneously. By installing this program in a phone, users can simultaneously operate multiple Java programs. And if a service provider places the appropriate software in its servers, users can access sites while simultaneously using various programs. The phone's processing speed and memory determine the number of applications that can be simultaneously run and the response time with these applications.

Another solution is to change the operating systems (OS) that are used in phones. In addition to facilitating multi-tasking, this will also facilitate file management including the saving of Java programs, photos, ringing tones, maps, and business data and their transfer to newly purchased phones. Symbian was supposed to provide phones with a standard operating system but support for Symbian is still uncertain. Most Japanese manufacturers use TRON or a proprietary version of TRON (originally developed for Japanese PCs in the 1980s) or Symbian's OS and the winning OS has still not yet been decided.

This is one place where Microsoft might be able to enter the mobile field. In addition to multi-tasking, the ability to easily exchange files between phones and PCs would be valuable for many users. On the other hand, Microsoft's solutions are based on Windows and they are currently too complex for the small screens, keyboards, processors, and memories in phones. This is a major reason why many industry observers believe that Mobimagic, the much much-ballyhooed joint venture between Microsoft and NTT DoCoMo, has produced few results in its three years of existence. And by the time it is possible to effectively apply Microsoft Window's solutions to the mobile phones, it is quite possible that other solutions will have already succeeded and created the necessary network effects to prevent Microsoft from succeeding in mobile phones.

STAGE 3 SYSTEMS: MAINTENANCE APPLICATIONS

Modern economies require continuous maintenance of items such as elevators, copy machines, computers, and power and transportation systems. Managing these engineers can be a complex and expensive business; ideally you would like to choose the engineer with the right skills and the right location and provide that person with the right information about the customer. Stage 3 mobile Intranet systems and GPS devices provide additional tools in the struggle to achieve these goals.

Japan Business Computer Corporation

Japan Business Computer Corporation (JBCC) provides maintenance for IBM computers. It has 16 branches, 74 offices, and maintenance contracts with 2000 companies. Its maintenance engineers make about 18,000 visits each month to its clients' offices.

Previously, an operator in the call center chose an engineer based on his or her best judgment and sent a message to that engineer's pager. The engineer then made a phone call from a pay or mobile phone to discuss the customer's problem with the operator. The operator explained the problem to the engineer and if it was determined that the engineer should be assigned to the customer, the operator verbally provided the engineer with additional information. The operator then called the customer and confirmed the visit. The engineer also called the operator when he arrived at the customer's site and when he completed the work and the operator updated the databases accordingly.

Mobile mail and other computer technology play a key role in the new process (see Table 9.2). When an operator receives a call from a customer, the operator chooses an engineer based on information about the available engineers in the computer database. The operators sends them mail that contains information about the customer both in the main body of the mail and in a URL that is included in the mail. The main body of the mail includes the firm's name, phone number, and

TABLE 9.2. JBCC's Maintenance Process

Process Step	Old Method	New Method	Comments
1. Customer contacts call center	Telephone	Telephone	
2. Call center chooses engineer	Best judgment	Computerdatabase	
3. Call center contacts engineer	Telephone	Mail	Mail and URL contains customer-related information and place to confirm visit
4. Call center confirms visit to customer	Telephone	Telephone	
5. Engineer updates call center on project status	Telephone	Mail	Update status in URL, which automatically updates database

address, and a place for the engineer to confirm his or her ability to visit the customer in a return mail message. The URLs, which make this application a Stage 3 system, provide links to maps, maintenance contracts, spare parts, and project status. The contract information includes the product manufacturing number and whether the product is still under guarantee. The links to the spare parts enable engineers to order parts, which are delivered via special courier.

Engineers update the project status as they complete the project by choosing from a list of choices in the URL. These choices include waiting for the next project, waiting for more information, in transit to next project, started task, finished task, and returning to the office. All of this is done with the push of one button, and engineers are not required to search through long menus.

The operators can access the project and engineer status information from PCs in the call center. Each engineer's schedule includes their own inputs and inputs by their immediate superior (e.g., meeting times). Engineers can also access their schedules on their mobile phones, and operators use these schedules to choose engineers for assignments.

The new system has reduced telecommunication costs, increased labor productivity, and increased customer responsiveness. Telecommunication costs were 37% lower than those of the previous year, which involved savings of more than one million yen ($U.S. 8,333) per month. Call center perators can handle more calls, engineers can solve more problems, and the engineers arrive at the customer's location faster than before JBCC introduced the new system.

JBCC is now considering the use of GPS, camera phones, and mobile groupware. GPS would enable operators to more effectively choose engineers by also taking into account their location. It is also planning to use camera phones to better discuss customer problems with engineers at the call center. Currently, a hardwire connection enables engineers at the call center to view information that is displayed on the customer's mainframe display. A camera phone makes it easier to relay information about other parts of the mainframe to the call center while maintenance engineers are discussing problems with engineers in the call center.

JBCC also wants to implement mobile groupware functions. It has already implemented an Internet-based groupware, and it would like to make this information available on the mobile phones of its customer engineers. It expects that such a system would improve communication between engineers, particularly through the sharing of engineers' schedules.

NEC

NEC's maintenance group introduced a mobile Intranet solution in September 2001 that is somewhat similar to JBCC's system. NEC's group has about 5000 engineers who service a variety of office equipment and computers. One of their main jobs is the replacement of faulty parts.

In the new system, operators use an integrated system of computers and phones to identify the customer and the history of this customer when a call comes in. Through discussions with the customer, they identify the problem and the needed parts and contact the closest regional service center. The closest service center prepares the parts and identifies an appropriate engineer using a computerized schedule and capabilities list. Like the JBCC example, the operator sends mail (see Table 9.3) to the engineer's phone and the engineer can access additional information and update their project status by clicking on one of the URLs included in the mail.

A unique aspect of this system is that the engineers can listen to a recording of the customer's call by clicking on a specific place in the

TABLE 9.3. Contents of Mail for NEC's Maintenance System

---

1) Operation summary
2) Appointment time
3) URL
4) Equipment type
5) Name of company and phone number
6) Part number for replacement
7) Problem description

---

mail; the particular function is called "phone tag" and it involves inter-active voice response. This function has significantly reduced the number of miscommunications between the call center and engineers. Previously problems were often incorrectly communicated to the engineers, resulting in incorrect diagnoses and unneeded visits. A critical aspect of the interactive voice response function is that because the system authenticates the phone when such a request is made, the recording cannot be accessed from other phones.

Other links in the mail messages provide access to other engineers and services - for example, companies that dispose of old parts. If the problem requires more time than expected or if the engineer can't understand the problem, the engineer can call a more experienced engineer whose phone number is also included in the mail. When engineers travel by public transportation, it is hard for them to carry the parts that were replaced. Thus the original mail message also contains a link to firms that dispose of these parts so that engineers can easily request the disposal of these parts.

The new system handles about 60,000 calls per month. In 70% of the cases, the engineer can respond (can or cannot do the job) within 3 minutes, which is much shorter than before[7]. Call center operators and engineers are able to handle more calls, and NEC was able to reduce telecommunication costs by more than 100 million yen per year, or 2068 yen ($17.20) per engineer per month.

STAGE 3 SYSTEMS: CONSTRUCTION

The United States is the leader in applying computer and Internet-based solutions to the construction industry including computer-aided design, online procurement, project management, and online auctions. The mobile phone provides an additional tool for the construction industry. In the short run, workers can use their phones to update process schedules, and in the long run they can access information about specifications, record tasks with the camera function, and keep track of materials with the GPS function.

One of the leaders in using both the PC and mobile-Internet in the construction industry in Japan is Sumitomo Construction. Sumitomo Construction first applied Internet tools to its home construction business where it builds over 10,000 custom homes per year. It builds these homes using a network of 430 offices, several thousand computers, and several thousand mobile phones. The offices are divided into sales and work offices. As of mid-2002, 2000 of the expected 5000 PCs had been installed and 4000 phones were being used.

Sales personal prepare a home's specifications on the PC while consulting with the customer in a sales office. They choose dimensions, colors, and other features, and Sumitomo Construction shares this information with suppliers from an early stage. While previously the specifications and schedule were sent three weeks in advance of the start of construction, this information is now shared with suppliers as early as three months in advance. The sales offices choose the starting and completion dates and the work offices create the detailed process schedules.

Sumitomo's construction workers use their mobile phones to update process schedules. These schedules are divided into 200 steps, 70% of which can be input from a mobile phone and the remaining 30% can only be input from a PC. Of the mobile inputs, carpenters input about 100 of them while other workers input about 40 of them. The work offices send mail to the carpenters each day at about 2 P.M. The workers click on the URL, and they indicate whether they have completed a specific step or steps. After completing a specific step, the next steps in the process are included in the following day's mail. Carpenters complete, on the average, 1.5 process steps per day.

Each night, the work offices look at the process status and update the schedules, which includes when specialists like plumbers and electricians should be at the construction site and when materials must be delivered. The plumbers and electricians work multiple sites and only spend a few consecutive days at one site. They bring tools and materials to the site and then return to their own offices where they can access these schedules on their PCs. While in the future the plumbers and electricians may also provide updates to the process schedules via mail, as of late 2002 the work office managers do this for them since the plumbers and electricians may only work a few days per month at a

Sumitomo construction site.

Sumitomo Construction is currently on track to reduce the average construction time from 112 to 90 days (or about) 20% and reduce the average construction cost by 5% through the better scheduling that the mobile phones provide. Previously, the site manager spent a lot of time checking the status of various tasks and then contacting the plumbers and electricians. There were often multiple days lost due to poor scheduling, and site managers spent a lot of time re-checking the status of multiple steps before deciding to proceed, particularly in the cases where multiple steps need to be completed before the next process step can be started. The data input on the mobile phone enables the site manager to have better data on process steps and to spend more time managing as opposed to collecting data.

In the future, Sumitomo Construction plans to use camera and location information. Camera phones can be used in place of digital cameras, which are already used to record critical tasks in large Japanese, European, and U.S. construction projects for quality and legal reasons. For example, many firms also use Web cams to monitor construction progress[8]. One example of using camera phones for such tasks can be found with Maeda Construction. It gave 180 of its construction workers camera phones so they can photograph tasks and insert them in reports using a Java-based application. NTT DoCoMo's i-area service also enables the rough location of the photograph to be recorded in the report. As high-resolution cameras (over two million pixels) become standard options on phones and GPS services become cheaper, these trends will likely accelerate.

STAGE 3 SYSTEMS: REPORT PREPARATION

In addition to construction and maintenance, many other jobs also require the preparation of short reports after visiting customers. These applications include home care, plumbing, and cable TV installation and repair. In addition, many nurses complete simple forms on their mobile phones after visiting patients in their homes.

Another large application includes temporary staff agencies that provide firms with temporary office workers. One of the big challenges for these companies is to stay in contact with employees who may rarely come to the office. And like NEC, JBCC, and other similar companies that fix office and other equipment, the temporary staff agencies would rather have their employees working at a customer's site than being at their office. For this reason, a number of Japanese companies have begun using mobile Intranets to support this communication.

For example, consider the case of Net Garage, which introduced a Java-based system in December 2002 that relies on an ASP service. While previously employees of Net Garage kept in touch with head-quarters via phone calls and faxes, they now input their working hours, working place, and work content via a Java program (that uses almost 100 kilobytes of memory if the working data are included). Employees can also access past work schedules and requested holidays on their phones. Naturally, headquarters staff can access the information input via the mobile phones on their PCs. The new system enabled the firm to reduce the number of telephone operators from three people to one and to respond much faster to changes such as sick days.

A variety of standard packages are now available for this applica-tion. Employees can access and update schedules, expense reports, and working hours. Furthermore, using location information from the clos-est base station or GPS, supervisors can monitor the locations of em-ployees on their PCs.

STAGE 3 SYSTEMS: SFA APPLICATIONS

The largest Stage 3 application may be SFA (Sales Force Automation). Sales is one of the oldest and still one of the largest professions in the world even in countries where B2Bs have seen significant growth such as in the United States. Whether these B2Bs eventually reduce the need for salespeople and if so where will they do this first are important questions that need to be addressed. However, these questions are not within the scope of this analysis, and instead this analysis assumes that sales will remain an important profession for the forseeable future dur-ing which many firms will be implementing systems that enable their salespeople to access sales related information from their mobile phones.

Boku-Undo

For example, consider Boku-undo, a small company with only 150 employees including 27 sales personnel but with a long history (it was founded in 1805). This company supplies small stationary stores with traditional Japanese brushes and inks. Previously, sales personnel placed orders via telephone and faxes where incorrect communication often occurred even with the faxes where product codes were not written correctly. Boku-undo considered the implementation of laptops but rejected this proposal due to their cost and weight.

Instead, Boku-undo chose to use mobile phones to access inventory information and place orders. Using a number of services from KDDI and its subsidiary Tsuka phone, it was able to introduce a system for several hundred thousand yen (several thousand dollars). The system includes KDDI's web-based inventory and ordering system. They introduced the system in April 2002; by July, sales personnel were checking inventory, and by November they were placing orders with their mobile phones. A name recognition system enabled sales personnel to easily input one of 2000 registered company names and access inventory or order about 3000 different items. Inventory can be checked in seven places and at four stages in the manufacturing process.

The results include reduced telecommunication and labor costs. By replacing phone calls and facsimiles with mobile Intranet accesses (and thus ASP and packet charges), Boku-undo was able to reduce average telecommunication costs per sales person from 15,000 yen ($120) to 10,000 yen ($80) per month resulting in more than one million yen ($8300) per year in savings. Coupled with reduced overtime for sending faxes and correcting mistakes, Boku-undo has realized savings of about three million yen ($25,000) per year.

Sony

In the summer of 2001, Sony began implementing an SFA system called e-mouse that relies on mobile mail. The initial cost was 50 million yen, or $420,000. By early 2002 about 1740 salespeople were using their mobile phones to check mail, schedules, wholesale prices, inventories, and sales figures. This number had risen to 2700 by mid-2003, 500 of which were using Java programs to access information. Sony is introducing a similar system for its maintenance employees.

The results are greater time spent visiting customers. While previously salespeople spent most of their mornings in the office doing administrative work and only visited companies in the afternoons, they now spend most of their work days visiting companies. They do their administrative work on their mobile phones between visits either in their cars, on public transportation such as trains, or in coffee shops.

Sony considered laptops, but their high cost, short battery times, and heavy weights caused Sony to choose mobile phones. Another reason for the advantage of the mobile phones over laptops is the difficulty of using laptops in combination with public transportation such as trains. They are heavy and easier to charge in a car than on a train. On the other hand, regardless of the method of transportation, the mobile phone is cheaper, easier to use in crowded places such as coffee shops, and, unlike laptop computers, is always on.

SUMMARY

The disruptive nature of the mobile phone is causing an unexpected set of business users and applications to emerge. Instead of firms using mobile phones to access existing ERP (Enterprise Resource Planning) and CRM (customer relationship management) systems, there is a different set of users, many of whom are doing their own systems integration and are the sources of the key innovations. The early applications include delivery, construction, maintenance, and sales, and the key technological trends include larger displays, increased processing power and network speeds, Java, and the effect of these on improving the phone's user interface.

Improvements in the user interface will most likely expand these business applications with the first step being Java. Many of the new application tools are based on Java and this will likely accelerate in the near future. Furthermore, faster processing speeds will lead to the greater use of 3D images including 3D displays of data, voice recognition, and other forms of new interfaces.

It will be interesting to see how general solutions, sometimes called dominant designs, emerge and evolve as the user interface is improved. Although many suppliers claim to offer standard solutions, most users will not buy a general-purpose package until they perceive that a general solution has emerged. If new user interfaces completely change the design of general-purpose packages, it is probably not in the best interests of users to quickly purchase such general-purpose packages. Alternatively, if suppliers of these systems are able to create an effective strategy for how to upgrade their packages (and installations) as the user interface evolves, general-purpose solutions might quickly emerge.

This will likely depend on the complexity of the application. Standard solutions will probably emerge much more quickly for Stage 1 and 2 systems and simple Stage 3 systems such as report preparation and taxis. In fact, most users already rely on such standard solutions for these applications.

On the other hand, it appears that the emergence of standard solutions for more complex Stage 3 applications such as maintenance, construction, and SFA will take longer to appear. Here users must integrate a number of mail, groupware, and other standard solutions including existing ERP and CRM systems. How the latter process will play out is still unclear. Will this necessary integration give existing ERP and CRM systems the opportunity to became leading suppliers of mobile solutions? Or will the firms that introduce these first mobile-based SFA systems like Sony become major suppliers of the software?

In maintenance, at least one of the two companies, JBCC, and a third company (Otsuka Shokai) that has implemented a similar system would like to sell their systems to other companies. JBCC modified software called "Service Alliance" that was originally created for laptops. U.S.-based ASTEA, a supplier of CRM software, developed Service Alliance for laptops so that maintenance workers could receive requests in the field and respond to them through a connection between the laptops and their mobile phones. JBCC modified the mail and other functions for the small size of the mobile phone. Otsuka Shokai is an office equipment and software supplier that is now trying to sell a general-purpose solution to maintenance groups. On the other hand, like the SFA systems, the integration of such maintenance system with existing information systems provides existing suppliers of ERP and CRM with an opportunity to become suppliers of the mobile software.

How these standard solutions become available outside of Japan is a more complex question. Most Japanese firms are waiting until mobile Internet services begin to grow in the United States or Europe before they begin trying to offer their standard solutions in these countries. Furthermore, the Western world may be slower than Japan and other Asian countries to use mobile phones in place of laptops to access their corporate intranets since public transportation plays a smaller role in these countries. But as Western firms begin to use mobile phones to access corporate Intranets, it is likely that Japan's lead in introducing such phone-based mobile Intranets will give them an advantage in the supplying of standard solutions for these mobile Intranets.

NOTES

[1] One study found that three-fourths of the first computers (implemented between 1944 and 1950) were developed by users. Knight, K., "A Study of Technological Innovation: the Evolution of Digital Computers," doctoral dissertation, Carnegie Institute of Technology, Pittsburgh, 1963.
[2] For example, see von Hippel, E., *The Sources of Innovation*, Oxford University Press, 1988.
[3] Data usage as a percent of revenues for corporate users was less than 10% in 2002, or far less than for regular mobile Internet users.
[4] See "UPS Case Study," AIR2Web (http://www.air2Web.com/wireless_success.jsp
[5] For example, Swedish taxi companies use car navigation systems and wireless techniques to monitor the location of taxis. These systems also include security features for taxis in trouble.
[6] When a call comes in, the page is automatically saved and reappears after the phone call is terminated.
[7] Since there was no way to measure the time it took for engineers to respond to requests from the call

center before the new system was implemented, hard data do not exist for the actual reduction in response time.

[8] I am indebted to Jari Veijalainen for this insight.

Chapter 10

Platform Strategy

The compatibility that is needed between different components in a network product like the mobile Internet often causes platforms to emerge that ensure this compatibility. Wintel computers play this role in PCs, Ethernet plays this role in LANs, and a whole host of standards play this role in the PC Internet. And the fact that the sponsors of these platforms often receive high profits makes the subject of platforms very interesting. Which technologies and firms will play this role in the mobile Internet? While no single firm will probably dominate the mobile Internet to the extent that Microsoft dominates, some firms will do much better than others.

The conventional wisdom is that it is a battle between Nokia and Microsoft. Nokia dominates the mobile phone industry while Microsoft dominates the PC industry. As the two industries converge, the two giants collide. Or so the story goes[1]. Others argue that NTT DoCoMo or Vodafone will be the platform leaders.

This concluding chapter argues that the issue is far more complex than this. It first looks at the power brokers in the mobile Internet. Here it is important to remember that the previous generation of leading firms often does not win in the next generation of technology, particularly when the next generation involves disruptive technologies. But in their selection of suppliers and partners, these power brokers can often determine who wins in the next generation of technology. Second, this chapter looks at both the conventional and unconventional views of platform management, of which technology plays a key role of in determining the winning platforms in the latter view. Here the importance of technology goes beyond creating a good platform and requires a key understanding of the technological trajectories. Third it applies the key technological trajectories and power brokers to the case of the mobile Internet in two concluding sections.

POWER BROKERS

Microsoft

Microsoft controls the PC world largely because IBM, which was the power broker in the previous generation of technology, chose Microsoft to be its supplier of OS software and managed to make its PC the industry standard. Users wanted applications such as games and spreadsheet software, and IBM's computers quickly outdistanced Apple in providing a large variety of this software. For its part, Microsoft supplied a decent operating system and figured out how to control the interfaces between the operating system and applications software.

Mobile Internet service providers should remember that IBM's decision to provide an open platform was its only chance of success. Furthermore, this strategy is even more relevant in today's world where consumers expect more variety than ever and the sources of this variety are more likely to be an unknown startup than a large firm's research division. Service providers that do not attract a large number of third-party content and technology providers will likely end up in the same position as Apple Computer did. On the other hand, IBM's mistake was to cede control of critical interfaces to suppliers like Microsoft, a mistake that service providers and phone manufacturers should clearly try to avoid.

In thinking about Microsoft's chances in the mobile Internet, it is useful to compare Microsoft's performance in browsers and PDAs; browsers were not a disruptive technology, whereas PDAs were. It was easy for Microsoft to add browsers to its Windows platform while maintaining compatibility with other application software. This has so far not been the case with PDAs where the smaller processing power in PDAs makes it difficult for Microsoft to maintain compatibility between Windows and Windows CE. Of course, the longer Palm takes to create a large market for PDAs and thus realize the benefits of the network effects associated with software and contents, the faster the processors in PDAs become. And the faster the processors become, the greater the chance that Microsoft will be able to use compatibility between Windows and Windows CE as a weapon in the PDA market.

The mobile phone industry is also disruptive technology for Microsoft, and thus Microsoft's efforts to sell software that is compatible with Windows to mobile phone manufacturers will probably have even less of a chance with phones than it did with PDAs. Furthermore, unlike PDAs, the mobile phone has already become a mass market with more than 400 million sold in 2002 or more than 30 times the number of PDAs sold in 2002. Microsoft can't afford to wait until phones have enough processing for their Windows software. It needs to create useful software for the mobile phone even if this requires the sacrifice of compatibility with its Windows and Windows CE software. Furthermore, if entertainment applications provide the road from which the winning platform is created for the mobile phones, Microsoft may need to address a completely new set of users.

Nokia

Nokia is the leading phone manufacturer in the world, with a share double that of its leading rival, and it has a strong brand name. It has created effective platform strategies in both the GSM market and at the global level where its global development capabilities enable it to develop more phones for the same development cost as its rivals[2]. These capabilities make it hard for other firms to take share from Nokia, and Nokia's high share enables it to directly determine the technologies that are put in almost 40% of the phones sold in the world and indirectly influence a large percentage of the other phones. The question is whether Nokia will be able to obtain these technologies under favorable conditions, and in this sense it has two challenges.

First, Nokia has so far not been a leader in defining future mobile Internet services, contents, or technologies. Although its phones are used in services such as SMS, MMS, and Vodafone Live!, it is a merely a supplier in these services as opposed to a creator. Nokia's main emphasis on defining the future of the mobile Internet appears to be the Open Mobile Alliance (OMA), which many argue is a rehash of the WAP Forum in that the OMA standard setting activities are not connected with the market (at least as of mid-2003). Nokia also does not participate in the Japanese mobile Internet where most of the key technologies are first tested.

Thus, although Nokia has the power to choose the technologies that will go into its phone, it could end up like IBM did in the PC industry and be forced to rely on other firms for the key technologies. Furthermore, if these new technologies fundamentally change the design tradeoffs in the mobile phone, other firms - for example, Japanese firms - may develop an advantage in phone design over Nokia. Such changes in the design tradeoffs have already caused significant changes in market share in Japan[3], and similar things may occur outside of Japan.

Second, the mobile Internet is also disruptive for Nokia's current business model that focuses on business users. Business users aren't the lead users in the mobile Internet, young people are. Nokia appears to have partly learned this in the messaging area and is now placing more emphasis on SMS and MMS. However, the extent to which Nokia will change its focus to young users and the long-term effect of this change on Nokia's overall business model is still unclear. In any case, Nokia's ability to control the evolution of MMS and its transition to the mobile Internet will have a large effect on Nokia's ability to be the dominant player in the mobile Internet. As argued above, if the mobile Internet significantly changes the critical design tradeoffs within phones, Nokia may be at a significant disadvantage with respect to its Japanese rivals.

NTT DoCoMo

NTT DoCoMo is the leading service provider in Japan, with almost 60% of the mobile phone and mobile Internet users. Its early creation of positive feedback between users, content providers, and technology suppliers has led to higher packet charges per subscriber and more support from content and technology suppliers than its competitors. It also determines the specifications for its phones, which makes NTT DoCoMo the major power broker in the Japanese mobile phone market. However, NTT DoCoMo faces two problems.

First, NTT DoCoMo has very little power and apparently capabilities outside of Japan. Its investments in foreign operators such as KPN, Hutchison Telecom, and AT&T have not done well, and its i-mode services in Europe have not done as well as expected. Given NTT DoCoMo's lead in the mobile Internet in 2000, it should have been able to help its partners introduce successful services faster than it has done.

One reason for the lower than expected performance of i-mode in Europe has been NTT DoCoMo's failure to convince a larger number of phone manufacturers to support the i-mode standard in Europe. It took more than 18 months from the announced alliance with KPN to introduce i-mode services in Holland in the spring of 2002. Eighteen months later there were only two phones available for European i-mode services.

This problem reflects the domestic nature of i-mode's network effects. By being the service provider to create positive feedback between content providers, users, and phone manufacturers in the Japanese market, it has created a virtual cycle of success that is very difficult to break. However, transferring these network effects and the virtual cycle of success to Europe or elsewhere is very difficult since European users do not necessarily want Japanese contents or even phones. NTT DoCoMo could have focused on using its knowledge of the relevant contents, business models, and technology to create a service more compatible with European technologies like SMS and WAP since this would probably have encouraged European phone manufacturers to support i-mode. Instead, NTT DoCoMo is attempting to transfer its i-mode system without changes to Europe and the United States.

A second problem for NTT DoCoMo is that it is becoming even more difficult for NTT DoCoMo to control the specifications in the Japanese mobile phones. As the number of key technologies widens and their complexity deepens, it must rely more on technology suppliers for advice on the specifications and gradually it is beginning to merely approve as opposed to create these specifications. For example, it relies on advice from firms such as K-Laboratories and Connect for Java, Hitachi, Intel, and TI for application processors, HI Corporation for 3D imaging techniques, Link Evolution for infrared technologies, and Sony's Bit Wallet and JR (Japan Railways) for non-contact smart cards. Like IBM and the PC, it is possible that NTT DoCoMo will lose control of these specifications, particularly because these technology suppliers work with all the service providers in Japan. Furthermore, many of these technology suppliers are working closely with foreign-service providers, and the linkage between the success of i-mode and these suppliers in Europe is becoming weaker all the time.

Vodafone

Vodafone is the leading service provider in the world with more than 100 million subscribers in 28 countries (as of May 2003). This enables Vodafone to strongly influence the future of the global mobile phone industry. For example, the greater success of Vodafone Live! when compared to the European i-mode services has caused many people to question the potential for success of i-mode in Europe. Looking backwards, if Vodafone had placed its full power behind the creation of an open WAP standard that emphasized entertainment contents and micro-payment systems, a different outcome would have likely emerged.

Vodafone's weakness is that it may not have the engineering capability necessary to define all the standards needed to support a successful platform. Its ownership of J-Phone, which is Japan's third largest service provider, undoubtedly helped Vodafone create Vodafone Live! and will probably continue to play a key role in making Vodafone Live! successful. However, Vodafone is trying to do things at a much more accelerated pace than J-Phone or NTT DoCoMo did, and it has far fewer engineers than NTT DoCoMo does. This is why many observers believe there are still a large number of undefined standards that are leading to large differences in how handsets access contents. These kinds of problems can eliminate Vodafone's advantages in the number of its handset suppliers for the Vodafone Live! services because content providers may only optimize their content for a few handsets.

A second and related challenge for Vodafone is managing alliances. Vodafone will probably manage alliances with European manufacturers better than NTT DoCoMo given its greater experience with them. On the other hand, the growing number of key technologies and firms will make these alliances more difficult to manage; even NTT DoCoMo is having trouble staying in control of these alliances. Furthermore, Vodafone will probably have to license Vodafone Live! to other service providers in order to make Vodafone Live! a winning platform, and as of September 2003, it was unclear how it would try to do this.

CONVENTIONAL PLATFORM MANAGEMENT

The conventional view of platform management can seduce these power brokers into complacency. The conventional view is that the design of the architecture and the alliances that firms make with others are the critical issues in platform management. Firms need to make decisions about the degree of modularity, the degree of openness of the interfaces to the platform, and how much information about the platform and its interfaces to disclose to outside firms. Firms also need to design their organizations and choose alliances that support the architectural decisions.

I do not dispute the importance of these decisions. Rather, I believe that the conventional viewpoint underestimates the importance of technology, which is much harder for power brokers to control than its alliances and architectures. Not only are successful platforms often technologically the best platforms[4], the key technological trajectories change the design tradeoffs and thus influence the emergence of so-called dominant designs. A dominant design is a single overall architecture that establishes dominance in a product class. Successful platforms can be considered part of or in some cases equivalent to dominant designs; but they should also be flexible enough to accommodate changes in the dominant design, which are driven by the technological trajectories.

These technological trajectories and their effect on dominant designs can threaten the power brokers since the power brokers often do not develop early experience with them and thus do not understand how they change the design tradeoffs and thus dominant designs. And without an understanding of these design tradeoffs and dominant designs, they cannot make good decisions in their chose of architecture or alliances. In the end, they might be forced to make alliances in which they operate from a position of weakness as IBM did with Microsoft and

TABLE 10.1. Effects of Technological Trajectories on Dominant Designs and Winning Platforms

Industry	Key Trajectories	Effect on Dominant Design and Winning Platform
Discrete transistors	Manufacturing Processes	Lack of improvement in frequency characteristics caused change from germanium to silicon.
Integrated circuits	Manufacturing processes	Lack of improvement in power consumption/heat dissipation caused change from bipolar to MOS and later CMOS.
NC controls for machine tools	Stepper motors	Improvements caused change from closed-loop to open-loop dominant design.
Electronic calculators	MOS ICs, displays	Improvements caused MOS ICs and LCDs to become part of dominant design.
Digital watches	CMOS ICs and displays	Improvements caused CMOS ICs and LCDs to become part of dominant design.
LCDs	Manufacturing processes	Inability to make color displays led to change to active matrix displays.
Mobile phones	ICs	Improvements caused change from analog to digital systems.

Source: Funk, J., "The Origins of New Industries," Hitotsubashi University, Institute of Innovation Research, WP#3-05

Intel.

KEY ROLE OF TECHNOLOGICAL TRAJECTORIES

Table 10.1 summarizes the effect of several technological trajectories on the dominant designs in several industries. Increased integration in integrated circuits led to problems with heat dissipation and thus changes in the dominant design from bipolar to MOS and later CMOS circuits. Much cheaper open-loop based controls replaced most closed-loop controls in machine tools as the performance of stepper motors improved and the feedback capability of closed-loop controls became less important. Markets for electronic calculators and digital watches did not grow until it became possible to use MOS and CMOS integrated circuits, respectively, in these products. The inability to make color displays caused a change from passive to active matrix LCDs. Improved integrated circuits enabled a move from analog to digital phones and a more efficient use of the frequency spectrum.

In each of the cases summarized in Table 10.1, the success of a new dominant design brought a new set of winners. Although all of the supporters of the old dominant design did not lose, many did partly because they misunderstood the key trajectories. And within the sup-

TABLE 10.2. Technological Trajectories, Dominant Designs, and Platforms: the cases of network products

Industry	Key Trajectories	Effect on Dominant Design and Winning Platform
Video players	Magnetic tape density	Improvements led to simpler designs and eventually longer playing time.
PCs	Processor Speeds	Allowed a new set of software suppliers to dominant the market.
LANs	Integrated Circuits	Improvements caused change to star-based topology.
PDAs	Processing Speeds	Will faster processing speeds enable Windows CE to catch up with the Palm OS?

Source: Funk, J., "The Origins of New Industries," Hitotsubashi University, Institute of Innovation Research, WP#3-05

porters of the new dominant design, it was only a few firms that did really well. These big winners, including TI in silicon transistors and integrated circuits, Fanuc in open-loop controls, Sharp in electronic calculators, Seiko in watches, and Sharp in LCDs, understood the trajectories and their implications for the evolution of the platform better than others.

The existence of network effects can further magnify these changes in market share. Table 10.2 summarizes the effect of technological trajectories on the dominant design several industries that involve network effects. Ampex dominated the broadcast market for video players from the early 1950s until the 1970s when improvements in magnetic tape density enabled Sony and Matsushita to begin selling a much simpler design to institutions like hospitals and corporations. And as further improvements in magnetic density were realized, longer playing time became a key method of competition and was to some extent a reason for the success of VHS over Beta. Ampex and Sony were both forced to exit the market.

Similarly, dramatic improvements in microprocessor performance was one reason why the IBM PC managed to become the dominant design, although it was released more than six years after the first PC was released by Altair in January 1975. While a different result might have emerged if Apple had pursued a more effective and open strategy, the dramatic improvements in microprocessor technology is another reason for the long time it took for a standard to emerge. And the greater processing capability of the IBM PC enabled a more sophisticated set of word processing, spreadsheet, and database programs to become the industry standards, thus driving a number of software producers out of business[5].

A key challenge for firms is to understand whether a critical mass will emerge for a dominant design before the technological trajectories change the design tradeoffs that underlay the dominant design and thus change the design. A few years ago, most people believed that Palm had developed a sufficient user base to make the Palm Pilot a dominant design. But with faster processors and a slowdown in PDA growth, it appears that the question is open again and Microsoft's CE is experiencing increasing sales.

THE EVOLUTION OF MOBILE PLATFORMS IN JAPAN

This book has described a number of trajectories that are propelling the mobile Internet forward and how they are doing this. Two of the key trajectories are faster network and faster processing speeds, both of which are highly interdependent. Both are needed to handle music, video, Java, and other applications.

Both will also influence the user interface, which is the key interface in the mobile phone. While the interface between the OS and applications software played this role in the PC and determined the winning platforms, contents and browsers for obtaining those contents are playing this role in the mobile Internet. Users subscribe to services that have or are believed will have lots of contents and phones. Similarly, phone manufacturers and content providers consider the number of subscribers and the actions of each other when they decide which services to support.

If it were possible to assume no future changes to the user interface, I would argue that the mobile Internet is a battle between i-mode, Vodafone Live!, and perhaps OMA in the United States, Europe, and other key markets outside of Japan. For many firms, including the suppliers of phones, browsers, and other technology, this is an important battle. But I believe that it is only easier to describe but not necessarily more important than the battle to create a new user interface, which is the largest bottleneck to an expansion of applications.

The first change will be from c-HTML to Java or an equivalent technology. Sun will be a likely winner, but it is likely that a more highly defined version or versions of a Java platform will emerge as part of the overall dominant design for the mobile Internet. This could be in the form of standard Java programs or standard components from which programs can be made. Due to restrictions in memory, processing speed, and network speeds, it is likely that these standards will emerge and many of them will be preloaded just as PCs, game platforms, and other products contain preloaded software.

Network effects will favor those programs or components that are able to establish an early installed base. The more such programs or components are used by consumers and content providers, the more likely that other content providers will want to base their contents on these programs or components and phone manufacturers will want to place them in their phones. Firms can influence the choice of platform by promoting their programs or components. Service providers or phone manufacturers can include them in their phones, and technology providers such as K-Labs or Connect can sell their programs or components cheaply.

However, it is likely that Java or a similar program will not be the last change in the user interface. It is quite possible that faster processing speeds will drive a number of changes in the user interface, each of which may change the platform and the winning firms. As processing speeds pass 500 MHz, 3D imaging techniques might become the new user interface for a wide variety of contents. And as processing speeds increase even further, voice recognition, virtual reality, holograms, and other still unknown technologies might also become the basis for new user interfaces and thus form the basis of future platforms.

THE EVOLUTION OF THE MOBILE PLATFORM IN THE WEST

How will the evolution of the mobile platform in Japan affect the rest of the world? Certainly there will continue to be competition between i-mode, Vodafone Live!, and perhaps OMA. However, there will also be competition between a host of other technologies such as Java, 3D imaging techniques, and voice recognition that will be played out beneath the surface as suppliers attempt to sell their version of these technologies to the service providers and phone manufacturers. This competition will occur initially in the area of entertainment contents but will likely also occur in other applications, and it is still unclear whether entertainment contents or other applications will determine the winning platforms.

In the PC, games created the critical mass in the industry but business applications determined the winning platforms. Similar things may appear in the mobile Internet where entertainment created the critical mass while the number of non-entertainment contents that are based on a specific program or component of that program may determine the winning platforms. It is quite likely that these programs and contents will come from Japan since this is where such development is currently occurring. Furthermore, also similar to the key role that played in the early days of the PC, the skills needed to develop such programs for text-based contents may come from the development of entertainment applications and thus so may the path and the winning firms.

The current emphasis on SMS and MMS by most Western service providers and even to some extent by Vodafone in Vodafone Live! may provide Japanese firms with an additional advantage. While SMS and MMS can handle many of the entertainment applications that are popular in the mobile Internet, they are less effective in handling browsing for information. If these non-entertainment contents determine the winning platforms, Japanese firms will have a distinct advantage as the West changes from SMS and MMS to mobile Internet services.

There will also be other important platforms, including ones that involve connections between phones and other devices, smart cards, payment systems, recognition/camera systems for URLs, bar code scanners, fingerprint and other biometric recognition systems, mobile Intranet applications, GPS solutions, and many infrastructure-related platforms. The winning platforms for these technologies will determine the future of many firms, particularly if some firms are able to make their technology a standard without completely opening their technology. Japan is taking a different path from the West in some of these areas, and in other areas Japanese firms are addressing issues that have not yet surfaced in the West. The West's focus on Bluetooth and Japan's focus on infrared technology for connections between phones and other devices is an example of the former. Japan is focused on infrared technology due to perceived power consumption problems with Bluetooth.

Like the platform issues associated with the user interface, there is likely to be simultaneous competition between alternative technologies both in Japan and in the West. Unlike the user interface, however, many of these technologies cannot be effectively used with SMS and MMS and require the implementation of a mobile Internet where complementary improvements in software, other hardware, and contents are needed for many of these technologies. Here it is possible that Japanese firms (or Western firms that participate in the Japanese mobile Internet) will dominate many of these platforms.

NOTES

[1] For example, see Nokia v Microsoft: The fight for digital dominance," *Economist*, November 21, 2002.
[2] For example, see Funk, J., *Global Competition Between and Within Standards: the case of mobile phones*, London: Palgrave, 2001.
[3] The success of i-mode in 1999 caused folding phones to become more popular than non-folding phones thus causing NEC, which was the main provider of folding phones, to replace Matsushita as the leader in the Japanese phone market.
[4] For example, see Liebowitz, S., *Re-Thinking the Network Economy*, NY: Amacom, 2002.
[5] Their developers were able to make Word Perfect and Lotus 123 more sophisticated and thus better than their competitors' products (Word Star and Visicalc) because they were designed around the higher processing speeds of the Intel microprocessor that IBM adopted.

Index